Daniel Farinotti

Estimating glacier ice thickness and snow accumulation

Daniel Farinotti

Estimating glacier ice thickness and snow accumulation

Simple methods for inferring the distribution of two variables of difficult access from readily available data

Südwestdeutscher Verlag für Hochschulschriften

Impressum/Imprint (nur für Deutschland/only for Germany)
Bibliografische Information der Deutschen Nationalbibliothek: Die Deutsche Nationalbibliothek verzeichnet diese Publikation in der Deutschen Nationalbibliografie; detaillierte bibliografische Daten sind im Internet über http://dnb.d-nb.de abrufbar.
Alle in diesem Buch genannten Marken und Produktnamen unterliegen warenzeichen-, marken- oder patentrechtlichem Schutz bzw. sind Warenzeichen oder eingetragene Warenzeichen der jeweiligen Inhaber. Die Wiedergabe von Marken, Produktnamen, Gebrauchsnamen, Handelsnamen, Warenbezeichnungen u.s.w. in diesem Werk berechtigt auch ohne besondere Kennzeichnung nicht zu der Annahme, dass solche Namen im Sinne der Warenzeichen- und Markenschutzgesetzgebung als frei zu betrachten wären und daher von jedermann benutzt werden dürften.

Verlag: Südwestdeutscher Verlag für Hochschulschriften GmbH & Co. KG
Dudweiler Landstr. 99, 66123 Saarbrücken, Deutschland
Telefon +49 681 37 20 271-1, Telefax +49 681 37 20 271-0
Email: info@svh-verlag.de

Zugl.: Zürich, ETH, Diss. No. 19268, 2010

Herstellung in Deutschland:
Schaltungsdienst Lange o.H.G., Berlin
Books on Demand GmbH, Norderstedt
Reha GmbH, Saarbrücken
Amazon Distribution GmbH, Leipzig
ISBN: 978-3-8381-2823-8

Imprint (only for USA, GB)
Bibliographic information published by the Deutsche Nationalbibliothek: The Deutsche Nationalbibliothek lists this publication in the Deutsche Nationalbibliografie; detailed bibliographic data are available in the Internet at http://dnb.d-nb.de.
Any brand names and product names mentioned in this book are subject to trademark, brand or patent protection and are trademarks or registered trademarks of their respective holders. The use of brand names, product names, common names, trade names, product descriptions etc. even without a particular marking in this works is in no way to be construed to mean that such names may be regarded as unrestricted in respect of trademark and brand protection legislation and could thus be used by anyone.

Publisher: Südwestdeutscher Verlag für Hochschulschriften GmbH & Co. KG
Dudweiler Landstr. 99, 66123 Saarbrücken, Germany
Phone +49 681 37 20 271-1, Fax +49 681 37 20 271-0
Email: info@svh-verlag.de

Printed in the U.S.A.
Printed in the U.K. by (see last page)
ISBN: 978-3-8381-2823-8

Copyright © 2011 by the author and Südwestdeutscher Verlag für Hochschulschriften GmbH & Co. KG and licensors
All rights reserved. Saarbrücken 2011

Preface

Snow and ice are important components of the cryosphere. The term 'cryosphere' comes from the ancient Greek word for 'freeze' or 'cold' and collectively describes all form of frozen water in the Earth system. Changes in climate alter the planet's snow and ice coverage. The increased pace of global warming has prompted concern about changes in the cryosphere. The role of snow and ice is particularly important in the formation of regional runoff and in its influence on global climate. Ensuring the future of water resources originating from snow and ice-melt poses a serious challenge to many nations. Snow and ice are known to have a significant impact on the runoff of alpine streams. During the summer months snow and ice meltwater may provide the main source of water for alpine valleys and dry lowlands. Knowledge of the present ice volume and the annual snow depth distribution and their expected changes in the coming decades due to climate warming is, thus, of great interest.

The author of this PhD thesis addresses the quantitative estimation of the ice thickness distribution throughout an entire glacier and the quantitative estimation of the snow accumulation within a catchment basin. Both variables are of fundamental importance for many applications, including glacier flow modelling and runoff forcasting and cannot easily be determined from remotely-sensed data. The author has developed two new methods to infer (1) the ice thickness distribution and, thus, the volume of a glacier from topographic data, a formulation of mass conservation and basics of ice flow mechanics, and (2) the distributed pattern of annual snow accumulation in an alpine watershed based on sequential topographic data obtained from terrestrial photography and snowmelt modelling. Moreover, method (1) was applied to the glaciers in Switzerland to provide, for the first time, a reliable estimate of the total ice volume present in this country.

Both methodologies can be considered as a strong foundation for and a fundamental contribution to future research and engineering work related to water resources in mountain regions.

Zürich, December 2010   Martin Funk

Contents

Preface i
Contents iii
Abstract v

1 Introduction   **1**
    1.1   The relevance of glaciers in a changing climate 1
    1.2   Monitoring and predicting glacier changes 2
    1.3   Objectives and structure of this thesis 4

2 Estimating an ice-thickness distribution   **5**
    2.1   Introduction 6
    2.2   Method 7
    2.3   Field data 10
    2.4   Results 10
    2.5   Discussion 17
    2.6   Conclusions 20

3 The glacier volume in the Swiss Alps   **23**
    3.1   Introduction 24
    3.2   Data 25
    3.3   Methods 27
    3.4   Results and discussion 31
    3.5   Conclusions 34

4 Inferring the snow accumulation distribution   **37**
    4.1   Introduction 38
    4.2   Study site and data 39
    4.3   Methods 40
    4.4   Results and discussion 46
    4.5   Conclusions 54

5 Conclusions and outlook   **57**

A Ice thickness measurements   **61**
    A.1   Overview 61
    A.2   Detailed information 62
    A.3   Maps 65

B Snow depth measurements on Dammagletscher   **79**

Bibliography   **81**

Abstract

Glaciers are prominent features in mountainous environments and with the ongoing climate change their importance has become aware to the wide public: Glaciers epitomize a healthy, untouched environment, are a key element of the water cycle, contribute largely to the current rate of sea-level rise and are amongst the most reliable climate indicators. Knowledge of recent behavior and future changes of glaciers is, thus, of great interest.

From the beginning of the last century, large efforts have been performed in order to build up and maintain networks for glacier monitoring. In this respect, the advent of remote-sensing techniques has contributed to fast progress. However, some fundamental quantities for the characterization of a glacier, such as the total ice volume, the distribution of the ice-thickness, the accumulation and the ablation, are difficult to determine from remotely-sensed data.

In this thesis, two new approaches are proposed for the inference of two distributed glacier variables which are generally difficult to estimate: ice-thickness and accumulation distribution. Both variables are of fundamental importance for many applications including ice-flow modeling, hydrological analysis or runoff forecasting.

A first method is proposed for inferring the ice-thickness distribution based on the principle of mass conservation and considerations on the ice-flow mechanics. Given a few parameters controlling the surface mass balance and the dynamics of the ice-flow, the ice-thickness distribution is derived from the characteristics of the surface topography. The result is thus, not only the total volume of a glacier, but also its spatial distribution. The method is able to integrate a-priori information about the ice thickness possibly available from radio-echo soundings or borehole measurements and is shown to perform with satisfactory accuracy.

The method is applied to a set of glaciers in the Swiss Alps and the results are used to calibrate a volume-area scaling relation. With the combination of the two approaches the total glacier ice volume present in the Swiss Alps in 1999 is estimated. The importance of large glaciers as contributors to the total volume is shown and an average mass-balance time-series is used to quantify the relative volume loss in the last decade. It is estimated that 12% of the 74 ± 9 km^3 of glacier ice present in the Swiss Alps in 1999, have melted until 2008. Out of this volume change, almost 30% was lost during the extraordinary warm summer 2003.

A second approach is proposed for inferring the distribution of snow accumulation by combining time-lapse photography and simple modeling. The parameters of a simple accumulation model, coupled to a distributed temperature-index melt model, are adjusted in an iterative procedure in order to reproduce the temporal evolution of the meltout-pattern observed during the ablation season. The small-scale variability is accounted for in a diagnostic way and lumped in a spatially distributed variable which summarizes all non-modeled snow redistribution processes. The comparison with in-situ snow-depth measurements shows that the achieved accuracy is in the same order as the results yielded by an inverse-distance interpolation scheme of the direct measurements. The influence of topographical variables is explored and local slope and curvature are shown to have a significant effect on the detected accumulation pattern.

Chapter 1

Introduction

1.1 The relevance of glaciers in a changing climate

At the latest since the *IPCC first assessment report* (IPCC, 1990), where the recession of mountain glaciers – meaning ice masses not in the Greenland and the Antarctic Ice Sheets – was used to provide qualitative support to the rise in global temperatures since the late 19th century, even the broad public has become aware of their importance: Glaciers are the symbol of an unviolated environment, are a key element in the water cycle, make a significant contribution to the current rate of sea-level rise and are visually and quantitatively amongst the most reliable indicators of climate change.

Even at low percentages of glacierization, catchment hydrology as well as the year-to-year variability of runoff, are significantly affected. The importance of glaciers in the water cycle is mainly due to the capability of snow and ice of storing water at many different time scales (Jansson et al., 2003). Long-term effects result from changes in the ice volume, are driven by the prevailing climate and manifest their influence in the order of decades. On the other hand, the characteristics of the interannual runoff is controlled by the seasonal cycle of snowcover build-up and depletion (e.g. Kasser, 1973; Østrem, 1973; Fountain and Tangborn, 1985). The year-to-year variability of runoff is lowest in catchments with moderate percentages of glacier cover and increases as glacier cover both decreases or increases (Fountain and Tangborn, 1985; Röthlisberger and Lang, 1987; Braithwaite and Olesen, 1988; Chen and Ohmura, 1990b).

The main difference between glacierized and glacier-free catchments is that the runoff from glacier-free basins is dominated by precipitation, whereas the runoff from glacierized ones is mainly controlled by the energy balance at the snow and ice surfaces (Lang, 1987; Chen and Ohmura, 1990b).

In Alpine regions such as Switzerland, parts of Austria, Italy or France, as well as in other mountain regions such as the Himalaya, the Andes or the Rocky Mountains, glacier changes play a major role. Especially during the summer months, the water supply of dry alpine valleys is often dependent from the streamflow contribution from snow and ice melt. And also the hydropower industry of mountainous regions relies often on melt water to fill the reservoirs. In a scenario of warming climate, water resources are expected to be altered in glacier-fed watersheds, and significant economical and social impacts are anticipated, especially in peripheral regions (Burlando et al., 2002).

Although mountain glaciers constitute only about 3% of the glacierized area on Earth and store less than 1% of the total ice volume, their contribution to the current rate of sea-level rise is not negligible (e.g Meier, 1984; Dyurgerov and Meier, 1997; Arendt et al., 2002;

Meier et al., 2007). The present total volume of mountain glaciers is estimated to be between 0.60 m (Radić and Hock, 2010) and 0.65 m of sea-level equivalent (Dyurgerov and Meier, 2005). Therefrom, about one quarter is estimated to have been lost by the end of the current century (Gregory and Oerlemans, 1998). This is about one third of the contribution which is expected from thermal expansion of sea water (e.g. Church et al., 2001).

The warming of the climate system is unequivocal (IPCC, 2007) and there is large consensus, that the tendency will not be inverted in the next decades. For the Swiss Alps, a further temperature increase of 1.8° C in winter and 2.7° C in summer has been projected until the year 2050 for the northern part of the Alps (Frei, 2007). This will cause major changes in the glaciers of our country (e.g. Zemp et al., 2006; Huss et al., 2008b) and force to face new challenges: new strategies for water- and hazard-management will be necessary in the near future.

1.2 Monitoring and predicting glacier changes

Since glaciers react sensitively to climatic variations, knowledge of recent changes and future behavior is of great interest. Different parameters can be used for characterizing a glacier. Length, area, surface-elevation, -slope and -aspect, as well as total volume or flow velocity are commonly used. For any kind of dynamical modeling, the initial geometry of a glacier has to be known. Information about the ice-thickness distribution is, in that sense, fundamental.

The temporal evolution of a glacier is mainly determined by the mass balance, which drives the ice-flow dynamics and reflects the prevailing climate conditions. Temporal changes in the mass balance result from changes in the accumulation and in the energy fluxes at the surface. Accumulation is mainly driven by precipitation and air temperature, whereas the energy fluxes at the surface steer the melting rates. While different approaches have been proposed for modeling melt processes and shown to perform satisfactorily (Hock, 2005), modeling snow accumulation still faces major difficulties (e.g. Dadic et al., 2010).

Permanently monitored glaciers and regularly updated inventories provide authoritative evidence for trends in the climate system on a global scale. First efforts for coordinating periodic observations of glacier changes were promoted at the end of the 19th century by F. G. S. Hall and were concretized in 1894 with the foundation of the *International Glaciological Commission*. The Commission, presided by F.-A. Forel, was in charge of "studying the variations of the dimensions of the current glaciers, in the different countries" (Forel, 1895) and was the forerunner of today's *World Glacier Monitoring Network (WGMS)*. The need for a worldwide inventory of perennial snow and ice masses was first considered during the *International Hydrological Decade* operated by UNESCO during 1965-1975. The *International Commission of Snow and Ice (ICSI* of the *International Association of Hydrological Sciences IAHS*) was asked to prepare guidelines on compiling glacier inventory data. These were produced in 1970 (UNESCO/IAHS, 1970) and refined in 1977 (Müller et al., 1977).

In Switzerland, the first systematic glaciological observations were carried out in the 19th century, when F. J. Hugi started his activities known as the *glacier campaigns* (1827-1831) and J. L. R. Agassiz investigated various processes on Unteraargletscher (Agassiz, 1840). F.-A. Forel was the promoter of the first yearly measurements of the front position of glaciers. These were started in the Canton Valais around 1880, in the context of analyses focusing on floods of Lake Geneva (de Saussure, 1880; Forel, 1892). These early studies can be regarded as the birth of the glacier monitoring programs in Switzerland.

Since then, large efforts have been performed by the *Swiss Glacier Monitoring Network* in

1.2. MONITORING AND PREDICTING GLACIER CHANGES

order to support continuous observations. Most of them have been collected in the *Yearbooks of the Cryospheric Commission of the Swiss Academy of Sciences* (Glaciological Reports, 2008). Recently Huss (2009) carried out a reanalysis of the mass balance data collected through large parts of the 20th century, including the time series of Claridenfirn, which started in 1914 and is the longest continuous mass balance time series worldwide.

The first comprehensive surveys, in which the morphology of the glaciers in Switzerland were assessed accurately, were conducted during 1833-1837 in the context of the compilation of the first official topographic map of Switzerland, the so called *Dufourkarte* (official name: *Topographische Karte der Schweiz*). The maps were published in 1845-1865 and had a scale of 1:100'000. An important revision of this epoch-making work, was carried out with the compilation of the *Siegfriedkarte* (official name: *Topographischer Atlas der Schweiz*), which was published during 1870-1926. The new maps had a scale of 1:50'000 in the Alpine regions, including glacierized regions. Jegerlehner (1902) analyzed those maps and gave the first comprehensive overview of the glacierized surfaces in the Swiss Alps. Although the work focused on the snow-line altitude of the glacierized regions, a compilation of morphological characteristics of each individual glacier, stating area, aspect and basic information about the hypsometry, as elevation range and mean altitude, was included. An updated version of these analysis was presented in the framework of the revision of the official maps of Switzerland. The revised *Landeskarte der Schweiz*, in scale 1:50'000, became the official cartography with the Federal law of 1935 and was at the basis of the glacier inventory presented by the *Eidgenössisches Amt für Wasserwirtschaft* in 1954. An exhaustive comparison of this data with those of Jegerlehner (1902) was performed by Mercanton (1958). In 1973, an updated glacier inventory for the Swiss Alps was presented by Müller et al. (1976). The work is known as the *Swiss Glacier Inventory 1973 (SGI1973)*. Since then, only one further update has been published, that is the inventory by Paul (2004), which is usually referred to as *SGI2000*. A revised version of *SGI2000*, elaborated from data collected in the year 2003 in the framework of the *LANDSAT* program of the *National Aeronautics and Space Administration (NASA)*, is currently in preparation (Paul, pers. comm. 2010) and is awaited with great interest.

In the last decades, remote-sensed data – meaning data which are acquired with techniques which do not require direct field access, such as aerial photography, satellite imagery or synthetic aperture radar (SAR) – have become an indispensable tool in monitoring programs, since large areas can be evaluated within one single survey. Recently, the *Global Climate Observing System (GCOS)* has called for the systematic monitoring of glaciers by satellites in support of the *UN Framework Convention on Climate Change*.

However, not all variables of interest can be remotely sensed. This applies, for instance, to mass balance as well as to ice-thickness distribution. It is thus not surprising, that, on a global scale, the spatial and temporal resolution of these data is fragmentary (Zemp et al., 2009).

In principle, area, length, surface elevation and flow velocity, can be determined rather easily from surface surveys. Although the data analysis requires some amount of work, the process of data acquisition can rely on established techniques, often based on remote sensing. In contrast, the determination of the ice volume of a glacier and its distribution is by far more laborious and, although airborne-based systems for measuring ice thickness were successfully experimented in the Alps (e.g. at Rhone-, Gorner- or Grosser Aletschgletscher), still requires direct field access. Current ice-thickness measuring techniques, as radio-echo soundings, borehole measurements or seismic-based approaches, provide only discrete measurements which are sparsely distributed in space. Thus, inter- and extrapolation approaches are required in order to provide spatially distributed estimates and considerable uncertainty is introduced in doing so.

The same is true for the determination of the components of the mass balance, i.e. accumulation and ablation. These components are usually determined at selected locations by means of the "glaciological method", that is by periodically reading stakes drilled into the ice for the ablation, and by carrying out manual snow-depth and -density measurements for the accumulation. This, of course, severely limits the spatial and temporal resolution of such measurements and requires, similarly to the determination of the ice thickness distribution, inter- and extrapolation techniques for getting information distributed homogeneously in space. Conversely, the ice-volume change of a glacier over a longer period corresponds to the "geodetic mass balance" and is easier to determine, as it can be computed by differencing two subsequent Digital Elevation Models (DEMs). Data for ice volume change in Switzerland exist for more than 50 glaciers (Bauder et al., 2007; Huss et al., 2010a,c) and are one of the most important datasets documenting the glacier evolution in the Alps throughout the last century. However, extracting distributed information about the mass-balance components from these data is still an unsolved task.

1.3 Objectives and structure of this thesis

This thesis develops new methods to infer two glacier variables which cannot be easily estimated from remote sensed data and whose ground-based estimation requires a considerable effort in terms of field work: the ice-thickness and the snow accumulation distribution. The performed work has been condensed into three peer-reviewed papers which were successfully published in three different journals (Farinotti et al., 2009a,b, 2010).

The inference of the ice-thickness distribution of a glacier from information derivable from characteristics of the surface topography is addressed first. A method based on glacier mass turnover and principles of ice-flow mechanics, mainly based on information which can be acquired from a surface DEM, is presented and successfully applied to four alpine glaciers in Switzerland for which direct ice-thickness measurements are available. The accuracy of the method is assessed and the potential for an application on a mountain-range scale is shown.

In the second part, the same method is applied for estimating the total ice volume present in the Swiss Alps and to quantify the relative volume loss that occurred during the last decade. The method is applied to every glacier in the Swiss Alps whose surface area is larger than 3 km^2 and the results are used to calibrate a volume-area scaling relation. Merging the two approaches, the total ice volume present in the Swiss Alps in the year 1999 is estimated to be 74±9 km^3. An average time series of mass-balance is then used in order to quantify the total volume loss in the last decade. The results indicate that 12% of the ice volume melted in the period 1999-2008 and that about 30% of this volume was lost during the extraordinarily warm summer 2003.

In the last part, a way to infer the snow accumulation distribution combining data collected with terrestrial time-lapse photography and simple modeling is presented. The method, based on an iterative adjustment of the parameters of a simple accumulation model, is developed and tested in a small basin of the central Swiss Alps (Dammagletscher catchment) but is easily applicable to any other region with a seasonally depleting snowcover. It is shown that the achieved accuracy is comparable to an inverse-distance interpolation of direct snow-depth measurements and that topographical parameters have a significant influence on the observed snow accumulation pattern.

The thesis is concluded with a "conclusion and outlook" chapter, in which the potential for further research and enhancement of the presented methods is discussed.

Chapter 2

A method to estimate ice volume and ice-thickness distribution of alpine glaciers

Citation: Farinotti D., M. Huss, A. Bauder, M. Funk and M. Truffer (2009). A method to estimate ice volume and ice-thickness distribution of alpine glaciers. *Journal of Glaciology*, 55 (191), 422–430.

ABSTRACT: Sound knowledge of the ice volume and the ice-thickness distribution of a glacier is essential for many glaciological applications. However, direct measurements of ice thickness are laborious, not feasible everywhere and necessarily restricted to a small number of glaciers. In this paper, we present a method to estimate the ice-thickness distribution and the total ice volume of alpine glaciers. This method is based on glacier mass turnover and principles of ice-flow mechanics. The required input data are the glacier surface topography, the glacier outline and a set of borders delineating different "ice-flow catchments". Three parameters describe the distribution of the "apparent mass balance", which is defined as the difference between the glacier surface mass balance and the rate of ice-thickness change, and two parameters the ice-flow dynamics. The method was developed and validated on four alpine glaciers located in Switzerland, for which the bedrock topography is partially known from radio-echo soundings. The ice thickness along 82 cross-profiles can be reproduced with an average deviation of about 25% between the calculated and the measured ice thickness. The cross-sectional areas differ by less than 20% on average. This shows the potential of the method for estimating the ice-thickness distribution of alpine glaciers without the use of direct measurements.

6　　　　　　　　　　　CHAPTER 2. ESTIMATING AN ICE-THICKNESS DISTRIBUTION

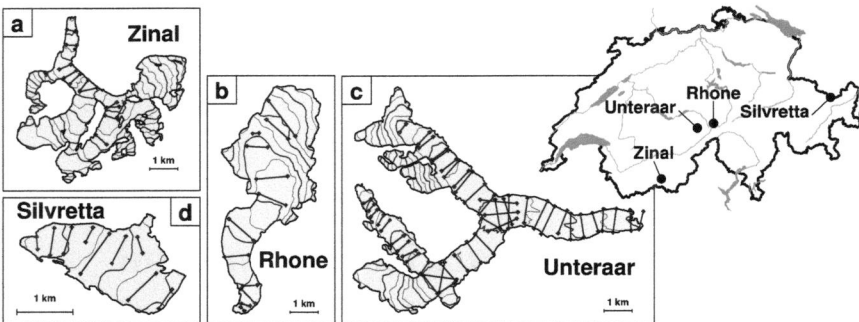

Figure 2.1: Location in Switzerland (map on the right) of the glaciers considered in this study (a-d). Glacier surface elevation is indicated by 100 m contours. Solid lines show profiles for which radio-echo soundings are available. Note that the inset of Silvrettagletscher has been enlarged by a factor of two in relation to other glaciers.

2.1 Introduction

A sound knowledge of the total ice volume and the ice-thickness distribution of a glacier is essential for many glaciological and hydrological applications. The total ice volume defines the amount of water stored by glaciers in a given catchment, and the ice-thickness distribution exerts an influence on the hydrological characteristics of the basin. Studies addressing the impact of climate change on the hydrology of high alpine catchments (e.g. Huss et al., 2008b) and most glacio-dynamical models (e.g. Hubbard et al., 1998) require the ice-thickness distribution as an initial condition.

Measuring the ice-thickness distribution of a glacier and deriving an estimate of its total volume is, however, not an easy task. Current ice-thickness measurement techniques, such as radio-echo sounding or borehole measurements, are expensive, laborious and difficult because of topographical constraints. Moreover, the determination of the ice volume of a glacier cannot be done directly, but is necessarily linked to inter- and extrapolation of direct (point) measurements.

For studies focusing on large samples of glaciers, it is necessary to develop alternative approaches which are based on readily available data sets. At present, the total ice volume of glaciers is often estimated using volume-area scaling relations (Erasov, 1968; Müller et al., 1976; Macheret and Zhuravlev, 1982; Chen and Ohmura, 1990a; Bahr et al., 1997). Several attempts have been made to infer the ice-thickness distribution on the basis of surface characteristics. These include applications of the shallow-ice approximation (Paterson, 1970; Haeberli and Hoelzle, 1995), or more complex procedures such as inverse methods based on modelling (Thorsteinsson et al., 2003; Raymond, 2007). The second category has focused mainly on ice sheets and ice streams (Collins, 1968; Gudmundsson et al., 2001).

We present a method for estimating both the overall ice volume and the ice-thickness distribution of alpine glaciers, based on mass turnover and principles of the ice-flow mechanics. The method was developed and validated on four alpine glaciers in Switzerland (Rhonegletscher, Silvrettagletscher, Unteraargletscher and Glacier de Zinal, Fig. 2.1), where ice thickness measurements are-available.

2.2 Method

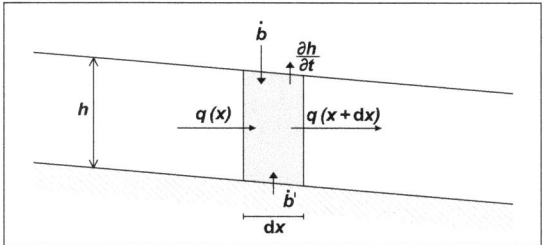

Figure 2.2: Schematic diagram of the concept of mass conservation for an ice element in a longitudinal glacier profile.

2.2 Method

The purpose of our method is to estimate ice-thickness distribution from a given glacier surface topography. According to the principle of mass conservation, the mass-balance distribution should be balanced by the ice flux-divergence and the resulting surface elevation change. The ice thickness can then be inferred from the ice fluxes. The theoretical background is as follows.

Consider a vertical column of ice with length dx and height h in a longitudinal glacier profile (Fig. 2.2). If the ice density ρ is constant and only plane strain is occurring, in the continuum limit of Figure 2.2, the mass conservation equation is of the form:

$$\frac{\partial h}{\partial t} = \dot{b} + \dot{b}' - \frac{\partial q}{\partial x}, \qquad (2.1)$$

where $\frac{\partial h}{\partial t}$ is the rate of ice-thickness change, $\dot{b}$ and $\dot{b}'$ the rates of mass gain or loss (mass balance) at the surface and the bed, respectively, and q the specific mass flux. In most glaciers $\dot{b}'$ is small compared to $\dot{b}$ (Paterson, 1994). In a more general, three-dimensional case the mass conservation equation can be written as:

$$\frac{\partial h}{\partial t} = \dot{b} - \nabla_{xy} \cdot \vec{q}, \qquad (2.2)$$

where $\nabla_{xy} \cdot \vec{q}$ is the ice flux divergence. Integrating Equation 2.2 over the glacier map domain Ω leads to:

$$\int_\Omega \frac{\partial h}{\partial t} d\Omega = \int_\Omega \dot{b} \, d\Omega, \qquad (2.3)$$

since one can write (using Gauss's law):

$$\int_\Omega \nabla_{xy} \cdot \vec{q} \, d\Omega = \int_{\partial \Omega} \vec{q} \, d\tilde{n} = 0, \qquad (2.4)$$

where $\vec{n}$ is the normal vector to the glacier outline $\partial \Omega$. In general, for a glacier, the spatial distribution of $\dot{b}$ and $\frac{\partial h}{\partial t}$ are unknown and difficult to estimate because of the complex spatial variability. We therefore introduce a new variable $\tilde{b}$ (m w.e. a^{-1}, the "dot" of the time derivative is omitted to simplify notation), which varies linearly with elevation and satisfies:

$$\int_\Omega \tilde{b} \, d\Omega = \int_\Omega \dot{b} \, d\Omega - \int_\Omega \frac{\partial h}{\partial t} d\Omega = \int_\Omega \left(\dot{b} - \frac{\partial h}{\partial t} \right) d\Omega = 0. \qquad (2.5)$$

The main advantage of estimating $\tilde{b}$ instead of $\dot{b}$ and $\frac{\partial h}{\partial t}$ separately, is that, under the condition in Equation 2.5, the integrated mass conservation equation (Eq. 2.3) is fulfilled without the need for information about the distribution of the surface mass balance and the spatial and temporal variation of the glacier surface elevation. If the given geometry corresponds to steady state, $\tilde{b}$ is the "actual" (i.e. the actual observable) glacier surface mass balance. In the following, $\tilde{b}$ will be referred to as "apparent mass balance".

The proposed method consists of estimating a distribution of the apparent mass balance, from which an ice flux for defined ice flow lines is computed. The flux is then converted into an ice thickness using an integrated form of Glen's flow law (Glen, 1955) and interpolated over the entire glacier. The resulting ice-thickness distribution is adjusted with a factor that accounts for the local surface slope. The result is an estimate of the ice-thickness distribution of the considered glacier.

The required inputs are (1) a digital elevation model (DEM) of the region; (2) a corresponding glacier outline; (3) a set of ice flowlines which determine the main ice-flow paths through the glacier; (4) a set of borders that confine the "ice-flow catchments" of the defined ice flow-lines (Fig. 2.3a); (5) three parameters defining the distribution of the apparent mass balance; and (6) two parameters describing the ice-flow dynamics.

Alpine valley glaciers are often composed of distinct branches flowing together into a main stream (see e.g. Fig. 2.1a and c). In this case, a division into different units, here called "ice-flow catchments", is necessary in order to represent the ice-flow dynamics. These catchments are digitized using topographic maps, aerial photographs or DEMs, tracking moraines or other geomorphological structures. The ice flowlines mainly correspond to the center of the ice-flow catchments.

The following assumptions are made: (1) the apparent mass-balance distribution can be described using two vertical mass-balance gradients: $d\tilde{b}/dz_{acc}$ for the accumulation area and $d\tilde{b}/dz_{abl}$ for the ablation area, and an equilibrium line altitude (ELA); (2) debris coverage reduces the apparent mass balance at a given location by a fixed percentage f_{debris} (Schuler et al., 2002); and (3) the glacier flow dynamics can be described by Glen's (1955) flow law.

The vertical mass balance gradients $d\tilde{b}/dz_{acc/abl}$ refer to the apparent mass balance $\tilde{b}$ and therefore do not necessarily correspond to the gradients of the actual mass balance. Glen's flow law is parametrized with the flow rate factor A, the exponent n and a correction factor C, which accounts for the valley shape (in analogy to the shape factor introduced by Nye, 1965), the basal sliding and the error arising from the approximation of the specific ice volume flux at the profile center.

The method can be subdivided into seven steps:

A) Using the given glacier surface topography and the defined vertical mass-balance gradients, the apparent mass-balance distribution is calculated for each ice-flow catchment. Starting with an estimated ELA z_0, the corresponding apparent mass balance $\tilde{b}_i$ for every grid cell i of elevation z_i is calculated using:

$$\tilde{b}_i = \begin{cases} (z_i - z_0) \cdot \left.\frac{d\tilde{b}}{dz}\right|_{abl} \cdot f_{debris} & \text{if } z_i \leq z_0 \\ (z_i - z_0) \cdot \left.\frac{d\tilde{b}}{dz}\right|_{acc} \cdot f_{debris} & \text{if } z_i > z_0 \end{cases} \quad (2.6)$$

with $f_{debris} = 1$ if the cell i is not debris-covered. The estimated ELA z_0 is then varied iteratively, until the condition of Equation 2.5 is met. The ELA which satisfies this criterion is an "apparent ELA" $\tilde{z}_0$, which does not necessarily correspond to the actual ELA. The procedure is repeated for every ice-flow catchment.

2.2. METHOD

B) With the apparent mass-balance distribution calculated in (A), the ice volume flux at each point of the ice flowlines is determined. This is done by cumulating the apparent mass balance $\tilde{b}_i$ of every grid cell i of the area that contributes to the ice volume flux at the considered point. The contributing area is approximated with the area located upstream of a line perpendicular to the ice flow line at the considered point (Fig. 2.3a).

C) The ice volume flux calculated in (B) is normalized with the local glacier width relevant for the ice discharge (here called "ice-discharge effective width") in order to obtain the mean specific ice volume flux $\bar{q}_i$. The ice-discharge effective width is determined along cross-sections perpendicular to the ice flow lines and is based on the surrounding topography (Fig. 2.3b). The local glacier width, determined by the intersection of the perpendicular line and the glacier outline, is therefore reduced to the width for which the slope of the ice surface does not exceed a given threshold α_{lim}. The error due to the approximation of the specific ice volume flux at the center of the cross profile (q_{center}) with the mean specific ice volume flux over the cross-profile ($\bar{q}$) is accounted for in the correction factor C.

D) When laminar flow is assumed in a parallel-sided slab glacier model (Paterson, 1994), the flow relation by Glen (1955) can be integrated and solved for the ice thickness. The ice thickness h_i at any point of the flow line with a mean specific ice volume flux $\bar{q}_i$ can then be calculated with the equation:

$$h_i = \sqrt[n+2]{\frac{\bar{q}_i}{2A} \cdot \frac{n+2}{(C\rho g \sin\bar{\alpha})^n}}. \qquad (2.7)$$

The contribution of basal sliding to the total flow speed is accounted for in the correction factor C. ρ is the ice density, g the acceleration of gravity and $\bar{\alpha}$ the mean surface slope along the considered ice flow line. In the following, the flow rate factor A is taken from the literature. Its uncertainties are transferred into parameter C. If ice-thickness measurements are available, the correction factor C can be calibrated. Otherwise, an estimation is required.

E) The ice thickness calculated in (D) is used as input for a spatial interpolation. The glacier outline is used as a boundary condition with zero ice thickness. The interpolation routine uses an inverse distance averaging technique, weighting the individual interpolation nodes with the inverse of the squared distance from the considered point.

F) To include the effect of the local surface slope α on the ice thickness, the ice thickness interpolated in (E) is multiplied by a factor proportional to $(\sin\alpha)^{\frac{n}{n+2}}$ at every grid cell. The factor is derived from Equation 2.7. The local surface slope is determined from a smoothed ice surface and is filtered with a lower slope limit α_0 to prevent an overestimation of ice thicknesses in very flat zones (as the factor tends to infinity when α goes to zero). The factor is normalized in order to preserve the previously calculated total ice volume.

G) Finally, the calculated ice thickness is smoothed with a two-dimensional discrete Gaussian filter of constant extension.

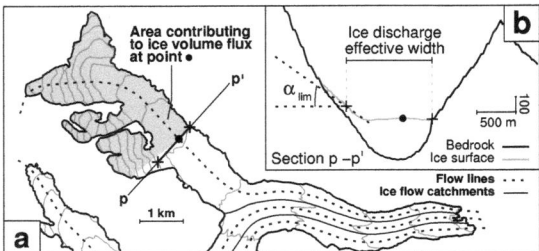

Figure 2.3: Determination of (a) the area contributing to the ice volume flux at a given point of an ice flowline, and (b) the ice-discharge effective width. Thin solid lines in (a) show the boundaries of the ice-flow catchments. The section p-p' is perpendicular to the ice flow line (dashed) at the considered location (dot). The ice-discharge effective width (boundaries marked by crosses) is determined using the glacier surface slope as a criterion, i.e. the boundaries are set where the slope exceeds a given threshold α_{lim}. The example refers to the northern tributary of Unteraargletscher.

The result is an estimate of the ice-thickness distribution of the entire glacier, which defines the overall ice volume. The bedrock topography can be calculated by subtracting the local ice thickness from the given glacier surface. A glacier surface smoothing is necessary to avoid the reproduction of small topographic surface features in the underlying bedrock.

2.3 Field data

The method is applied to four alpine glaciers located in Switzerland (Fig. 2.1). The selected glaciers comprise different glacier types. Unteraargletscher and Rhonegletscher are large valley glaciers. The former is divided into three main branches flowing together, and the latter has a compact geometry. Glacier de Zinal is a highly branched valley glacier of medium size and Silvrettagletscher is a small and compact mountain glacier. The ice-thickness distributions of the four glaciers are known along several cross-profiles from radio-echo soundings (Fig 2.1). The glacier surface topography is given by DEMs and glacier outlines for different years. The methods used for data acquisition are described by Bauder et al. (2003) and Bauder et al. (2007). The spatial resolution of the DEMs is 25 m for all glaciers. The available datasets are summarized in Table 2.1.

Unteraargletscher was chosen for the validation because of the large number of datasets available. These include radio-echo sounding ice-thickness measurements (Funk et al., 1994; Bauder et al., 2003), annual surface ice-flow speed measurements from 1924 to 2005 on cross-profiles in the ablation area (Flotron, 2007), and annual ice volume fluxes across profiles in the ablation area determined from surface velocity data (Huss et al., 2007).

2.4 Results

The method was applied to Rhonegletscher, Glacier de Zinal and Silvrettagletscher in order to calibrate the input parameters and to determine the sensitivity of the output with respect to the calibrated parameters. A validation was performed on Unteraargletscher.

2.4. RESULTS

Table 2.1: Available data sets. profiles (number of profiles with ice-thickness measurements); measurements (years in which the radio-echo sounding measurements were performed) and DEMs (years for which surface topography and glacier outlines are used).

Glacier	Profiles	Measurements	DEMs
Rhone	13	2003	1929, '80, '91, 2000
Silvretta	9	2007	2007
Unteraar	48	1990, '97, '98, 2000	2003
Zinal	12	2006, '07	2006

Table 2.2: Parameter values and units.

Parameter	Symbol	Value	Unit
Flow rate factor	A	$2.4 \cdot 10^{-15}$	$\frac{1}{(\text{kPa})^3 \text{s}}$
$d\tilde{b}/dz$ accumulation zone	$\frac{db}{dz}\big|_{\text{acc}}$	$0.5 \cdot 10^{-2}$	a^{-1}
$d\tilde{b}/dz$ ablation zone	$\frac{db}{dz}\big|_{\text{abl}}$	$0.9 \cdot 10^{-2}$	a^{-1}
Debris reduction factor	f_{debris}	0.5	—

For the sake of simplicity, the same values for the vertical mass-balance gradients $d\tilde{b}/dz_{\text{acc/abl}}$, the debris coverage reduction factor f_{debris} and the parameters of Glen's flow law A and n were used for all glaciers (Table 2.2). The only glacier-specific parameter is the correction factor C (Table 2.3).

The flow rate factor A was set to the value determined for alpine glaciers in previous studies (Hubbard et al., 1998; Gudmundsson, 1999). This value, obtained by fitting measured ice surface velocities to results of ice-flow models, is smaller by a factor of about two than that resulting from laboratory experiments (e.g. Paterson, 1994). For Rhonegletscher, Glacier de Zinal and Silvrettagletscher, the correction factor C was calibrated by fitting the calculated ice thicknesses to the radio-echo sounding measurements. For the validation on Unteraargletscher, C was set to the mean value calibrated for the considered valley glaciers, i.e. Rhonegletscher and Glacier de Zinal. The flow law exponent is $n = 3$. The values chosen for $d\tilde{b}/dz_{\text{acc/abl}}$ and f_{debris} are based on mass balance studies on the considered glaciers (Huss et al., 2007, 2008a). As an approximation, we assumed that the gradients of the apparent mass balance $d\tilde{b}/dz_{\text{acc/abl}}$ correspond to the gradients of the actual mass balance. The ice density was assumed to be $\rho = 900\,\text{kg}\,\text{m}^{-3}$ for all glaciers. The threshold slope α_{lim}, which defines the ice-discharge effective width, and the lower slope limit used for the surface slope filtering, were determined empirically and set at $\alpha_{\text{lim}} = 20°$ and $\alpha_0 = 5°$, respectively. For Rhonegletscher, Glacier de Zinal and Silvrettagletscher the mean absolute deviation between calculated and measured ice thicknesses is 26 m (corresponding to 19% of the measured ice thickness, standard error of estimate $SEE = 36\,\text{m}$) (Fig. 2.4a). The mean absolute deviation between calculated and measured cross section area is $1.6 \times 10^4\,\text{m}^2$ (15%, $SEE = 2.3 \times 10^4\,\text{m}^2$) (Fig. 2.4b). The sum of

Table 2.3: Glacier-specific values of the dimensionless correction factor C.

	Rhone	Silvretta	Unteraar	Zinal
C	0.45	0.80	0.65	0.85

CHAPTER 2. ESTIMATING AN ICE-THICKNESS DISTRIBUTION

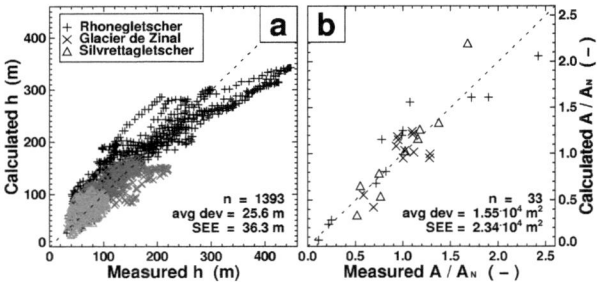

Figure 2.4: Comparison of calculated and measured (a) ice thickness h and (b) cross-sectional area A. For better visualization, the cross-sectional area is normalized with the mean measured area A_N. The statistics on bottom right refer to the whole ensemble of points. n: number of points, $avg\ dev$: average deviation, SEE: standard error of estimate.

Table 2.4: Key parameters for the four analyzed glaciers resulting from the method application. V_{Bahr} is the total ice volume determined using the volume-area scaling relation of Bahr et al. (1997).

Glacier	Year	Area (km²)	Volume (km³)	$\bar{h}$ (m)	h_{max} (m)	V_{Bahr} (km³)
Rhone	2000	15.12	1.97	130	345	1.33
Silvretta	2007	2.81	0.17	60	121	0.13
Unteraar	2003	22.71	4.00	176	369	2.31
Zinal	2006	13.41	0.82	61	174	1.13

all calculated cross section areas deviates by 3% from the sum of the measured cross section areas. On Unteraargletscher, used for validation, the absolute mean deviation between calculated and measured ice thickness is 39 m (20%, $SEE = 50$ m) (Fig. 2.5a). The absolute mean deviation between the calculated and the measured cross-sectional area is 2.6×10^4 m² (17%, $SEE = 3.3 \times 10^4$ m²) (Fig. 2.5b). The deviation of the sum of all cross-sectional areas amounts to 6%.

Table 2.4 lists the key parameters of the four analyzed glaciers. The ice volume determined using a volume-area scaling relation (Bahr et al., 1997) is reported for comparison purposes. Except for the highly branched Glacier de Zinal, the ice volume determined using the volume-area scaling relation is smaller than the volume calculated by our method (−37% on average). Numerical values characterizing the deviation between calculated and measured ice thicknesses for the individual glaciers are listed in Table 2.5. Figures 2.6-2.9 depict the calculated ice-thickness distribution; comparisons between calculated and measured bedrock topography along selected profiles are also.

The comprehensive data set available for Unteraargletscher allows two more aspects of the results to be validated. In the following, the ice volume fluxes and the surface flow velocities resulting from the presented method are compared with measurements and results of earlier studies.

Using the surface flow-speed measurements of Flotron (2007) and the ice-thickness distribution known from radio-echo soundings, Huss et al. (2007) calculated an annual ice volume flux (here referred to as "observed" ice volume flux) for different cross-profiles in the ablation

2.4. RESULTS

Table 2.5: Comparison between measured and calculated ice thickness and cross-sectional area. $\overline{|\Delta h|}$: average absolute deviation between measured and calculated ice thickness, SEE_h: standard error of estimate of $|\Delta h|$, $\overline{|\Delta A|}$: average absolute deviation between measured and calculated cross section area, SEE_A: standard error of estimate of $|\Delta A|$.

Glacier	$\overline{\|\Delta h\|}$ (m)	(%)	SEE_h (m)	$\overline{\|\Delta A\|}$ (10^4 m^2)	(%)	SEE_A (10^4 m^2)
Rhone	36	27.8	45	2.8	20.6	3.6
Silvretta	16	24.9	19	0.5	14.5	0.8
Unteraar	39	20.2	50	2.6	16.9	3.3
Zinal	22	19.8	28	1.0	16.1	1.2

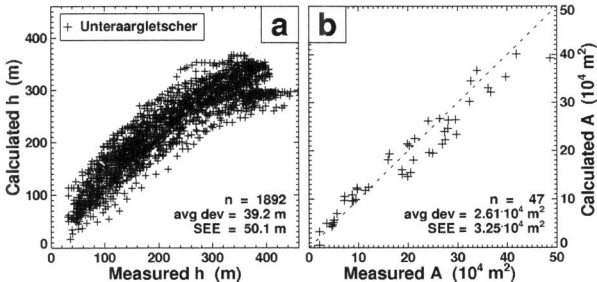

Figure 2.5: Comparison of calculated and measured (a) ice thickness h and (b) cross section area A for Unteraargletscher. n: number of points, *avg dev*: average deviation, SEE: standard error of estimate.

CHAPTER 2. ESTIMATING AN ICE-THICKNESS DISTRIBUTION

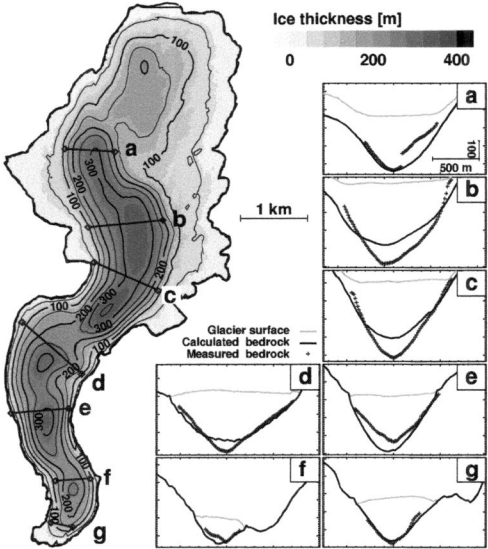

Figure 2.6: Calculated ice-thickness distribution of Rhonegletscher. The insets (a-g) show the marked cross-sections; all have the same vertical exaggeration.

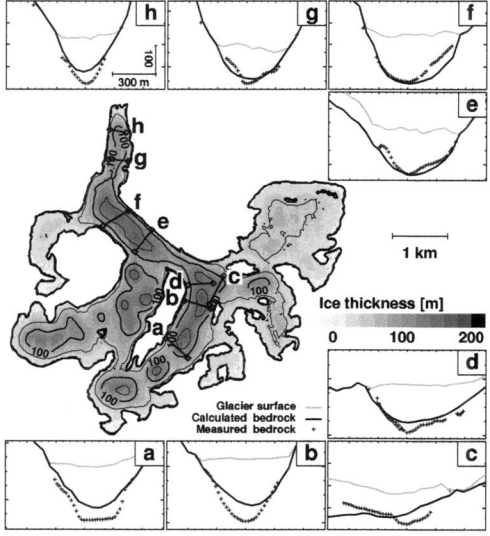

Figure 2.7: Calculated ice-thickness distribution of Glacier de Zinal. The insets (a-h) show the marked cross-sections; all have the same vertical exaggeration.

2.4. RESULTS

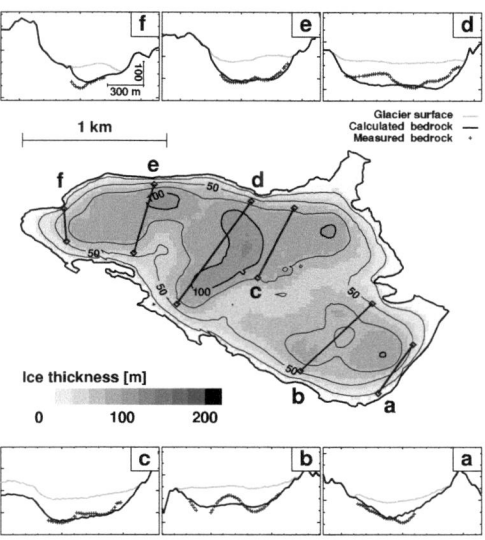

Figure 2.8: Calculated ice-thickness distribution of Silvrettagletscher. The insets (a-f) show the marked cross-sections; all have the same vertical exaggeration.

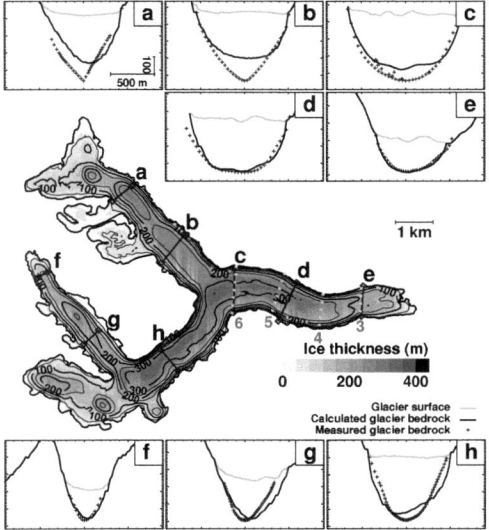

Figure 2.9: Calculated and measured ice thickness distribution of Unteraargletscher. The insets (a-h) show the marked cross-sections. All have the same vertical exaggeration. Ice-flow velocity measurements are available for the gray dashed profiles with numbers.

CHAPTER 2. ESTIMATING AN ICE-THICKNESS DISTRIBUTION

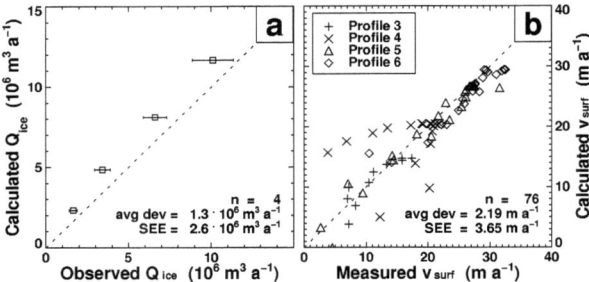

Figure 2.10: Comparison of (a) calculated and observed ice volume fluxes Q_{ice} and (b) calculated and measured surface ice-flow velocities v_{surf} for the four cross-profiles labelled with numbers in Figure 2.9. Observed ice volume fluxes represent mean values for the 1989 to 1998 period (bars corresponding to two standard deviations); measured ice flow velocities at surface refer to the year 2001.

area. The flux was determined by integrating the flow-speed field resulting from up to 30 surface velocity measurements per profile. The vertical distribution of the horizontal flow speed was calculated from measured surface velocities with an assumption of simple shearing (Paterson, 1994). These results are compared with the ice volume flux resulting from our method. The latter is determined by adding up the ice volume fluxes of the individual ice flowlines at the location where the flow lines are crossing the considered profile.

The observed ice volume flux can be reproduced within a factor of 1.3 for four cross-sections located at the tongue of Unteraargletscher (Fig. 2.9). The mean deviation between observed and calculated ice volume flux is $1.3 \times 10^6 \, \mathrm{m}^3 \mathrm{a}^{-1}$ ($SEE = 2.6 \times 10^6 \, \mathrm{m}^3 \mathrm{a}^{-1}$) and is almost constant for the individual profiles (Fig. 2.10a).

For the same cross sections considered above, an ice-flow velocity distribution was calculated using a model described in Sugiyama et al. (2007). The model calculates the horizontal flow-speed field along a given cross-section by solving the equations for balance of shear stress and Glen's flow law. Basal sliding is introduced by a linear relation between the sliding speed and the shear stress acting on the bed (Weertman, 1964; Lliboutry, 1979). The relation is given by the so-called "sliding coefficient". The glacier bedrock calculated with the presented method and the glacier surface topography were used as the boundary condition. For the flow rate factor A, the same value was chosen as in the calculation of the ice-thickness distribution (Table 2.2). The surface slope of each profile was derived from the DEM. The sliding coefficient was supposed to be constant across one single profile but was adjusted for each in a range of 30 to $90 \, \mathrm{m \, (a \, MPa)}^{-1}$ in order to match the respective maximum surface velocity. This leads to a basal sliding that accounts for 15 to 35% of the observed surface speed, which is slightly less than found by Gudmundsson et al. (1999) and Helbing (2005) by inclinometer measurements. The ice-flow velocities at the surface calculated using the model were then compared with the surface velocity measurements (Fig 2.10b). The average deviation is $2.2 \, \mathrm{m \, a}^{-1}$ ($SEE = 3.7 \, \mathrm{m \, a}^{-1}$).

2.5 Discussion

The accuracy of the method in estimating ice thicknesses is assessed by comparing calculated and measured quantities for Unteraargletscher, which was used for validation without tuning of the parameters. The accuracy of the ice-thickness estimation is, thus, about 20%, which corresponds to less than 40 m for this glacier. The cross section areas are estimated within 17%. The estimated accuracy is a measure for the resolution of the calculated glacier bedrock topographies. Bedrock features smaller than the estimated accuracy cannot be resolved and should not be further interpreted. This may partially limit the use of the results for applications where the resolution of small-scale features is required (e.g., 3D ice flow modelling of basal processes). In such a case, an assessment of the sensitivity with respect to the input bedrock geometry is recommended.

Our method is based on the determination of the ice volume fluxes of a glacier. The apparent mass balance $\widetilde{b}$, defined at any point as the difference between the surface mass balance $\dot{b}$ and the rate of thickness change $\frac{\partial h}{\partial t}$, is introduced in order to account for mass conservation. The main advantage of estimating $\widetilde{b}$, and not $\dot{b}$ and $\frac{\partial h}{\partial t}$ separately, is that the glacier-wide average of $\widetilde{b}$ is zero by definition (Eq. 2.5) and no steady-state assumption is therefore required. This is in contrast to other methods, such as volume-area scaling, that have an inherent steady-state assumption. An unique relationship between volume and area is only possible with glaciers in steady state. This assumption is often violated in today's climate with many glaciers out of equilibrium.

The distribution of $\widetilde{b}$ is described by two parameters, $d\widetilde{b}/dz_{acc}$ and $d\widetilde{b}/dz_{abl}$, defining the vertical gradient of $\widetilde{b}$ in the ablation and the accumulation zone, respectively. In general, the gradients $d\widetilde{b}/dz$ do not correspond to the gradients of the actual mass balance $d\dot{b}/dz$. A correspondence of the two is given only if the rate of ice thickness change $\frac{\partial h}{\partial t}$ is zero everywhere i.e. the glacier is in steady state. Thus, $d\dot{b}/dz$ is a good approximation of $d\widetilde{b}/dz$ when $\frac{\partial h}{\partial t}$ is small. According to Jóhannesson et al. (1989), $\frac{\partial h}{\partial t}$ is highest at the glacier tongue and becomes rapidly smaller as the distance from the tongue increases. Thus, in the accumulation zone, the approximation of $d\widetilde{b}/dz$ with $d\dot{b}/dz$ is justified in most cases and one can write $d\widetilde{b}/dz_{acc} \approx d\dot{b}/dz_{acc}$. A larger difference is expected between $d\widetilde{b}/dz_{abl}$ and $d\dot{b}/dz_{abl}$. The difference increases as the glacier is further out of equilibrium. In the ablation zone, $d\widetilde{b}/dz_{abl}$ is expected to be steeper than $d\dot{b}/dz_{abl}$. Figure 2.11 illustrates the result of the parametrization of $\widetilde{b}$ with $d\widetilde{b}/dz_{acc/abl}$ for the case of Rhonegletscher in the period 1991-2000. The distribution of the mass balance $\dot{b}$ with altitude was calculated based on direct measurements and a distributed mass-balance model (Huss et al., 2008a). The rate of thickness change $\frac{\partial h}{\partial t}$ was determined by differencing the DEMs of the years 1991 and 2000. Despite of the approximation of $d\widetilde{b}/dz_{abl}$ with $d\dot{b}/dz_{abl}$, the difference $\dot{b} - \frac{\partial h}{\partial t}$ and the estimated apparent mass balance $\widetilde{b}$ agree well in both, the accumulation and the ablation zone (Fig. 2.11b). Thus, in this case, the approximation $d\widetilde{b}/dz_{abl} \approx d\dot{b}/dz_{abl}$ is justified. The sensitivity of the present method with respect to the input surface topography was tested for Rhonegletscher. Bedrock topographies were calculated based on the surface topographies of the years 1929, 1980, 1991 and 2000. The comparison of the results showed that the difference between two bedrock topographies become larger as the time lag between the two considered input geometries increases. The mean specific net balance in the previous 20 years $\overline{b_{n,20}}$ is considered an indicator for how far the considered glacier geometry is from a steady state. Figure 2.12 shows the bedrock calculated for Rhonegletscher with the surface topogra-

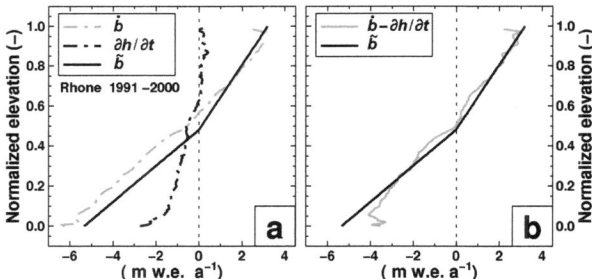

Figure 2.11: Altitudinal distribution of (a) modelled mass balance $\bar{b}$, observed rate of ice-thickness change $\frac{\partial h}{\partial t}$ (Huss et al., 2008a) and estimated apparent mass balance $\tilde{b}$ and (b) difference $\bar{b} - \frac{\partial h}{\partial t}$ and estimated $\tilde{b}$ for Rhonegletscher. The elevation range is normalized. Values are means over the period 1991-2000 and expressed in m w.e. a^{-1}.

phies for the years 2000 ($\bar{b}_{n,20} = -0.59$ m w.e. a^{-1}) and 1929 (+0.16 m w.e. a^{-1}) (Huss et al., 2008a). The mean deviation between the two bedrock topographies is $+22$ m, indicating that the bedrock calculated with the surface topography of the year 2000 has a higher average elevation. The mean deviation is smaller than the estimated accuracy. The largest differences occur in the ablation area (Fig. 2.12). This was expected due to the approximation of $d\tilde{b}/dz$ with $d\bar{b}/dz$. The sensitivity of the method with respect to the input parameters was tested on Rhonegletscher and Glacier de Zinal. The sensitivity is quantified in terms of relative change in the mean ice thickness for all parameters. For the mass balance gradients, the sensitivity is also expressed in terms of relative change in the calculated apparent ELA.

The flow rate factor A and the correction factor C are the most sensitive parameters affecting the calculated mean ice thickness (Fig. 2.13a). On average, a variation in C of 0.1 leads to a variation of 9% in the mean ice thickness. The parameters C and A are not independent of each other (Eq. 2.7). An increase in A by a factor two is equivalent to a decrease in C by 20%, and it reduces the mean ice thickness by 18% (Fig. 2.13a). The sensitivity of the two parameters diminish slightly towards larger values of A and C and are glacier-independent.

The sensitivity on the flow rate factor A and the correction factor C are intrinsic to the approach. Since uncertainties in A can be transferred to C, the flow rate factor A is set to values reported in the literature (e.g. Hubbard et al., 1998; Gudmundsson, 1999). The determination of C is more difficult, as it accounts for different approximations and uncertainties. These are: (1) the approximation of the shear stress distribution by a linear relation (Nye, 1965); (2) the approximation of the specific ice volume flux at the center of the profile q_{center} with the mean ice volume flux across the profile $\bar{q}$; (3) the influence of the basal sliding, assuming a linear relation with the deformation velocity (e.g. Gudmundsson, 1999); and (4) the uncertainties in the flow rate factor A.

Considering only approximation (1), the correction factor C depends on the shape of the cross-section only. In this case C assumes values between 0 (channel of infinite depth) and 1 (channel of infinite width) (Nye, 1965). For approximation (2), one can show that in the case of a channel with cylindrical shape, the approximation of q_{center} with $\bar{q}$ increases C by less than 10%. With approximation (3), a doubling of the surface speed velocity due to increased basal sliding, would require an increase in C by 25%. An uncertainty in A (approximation (4)) by a

2.5. DISCUSSION

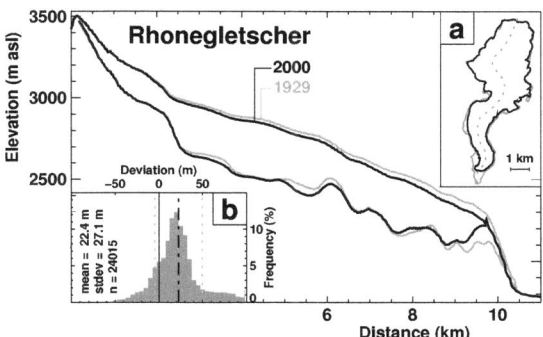

Figure 2.12: Comparison of the glacier bedrock along the central flowline of Rhonegletscher calculated using two different input geometries: (a) glacier extent for the years 2000 (black) and 1929 (gray) and central flowline (dashed); (b) distribution of the deviation between the two calculated bedrocks in the domain covered by both (24'024 grid cells, 15.0 km^2). The mean deviation (22.4 m) is marked by the black dot-dashed line, the range of two standard deviations ($\pm$ 27.1 m) by the gray dashed lines.

factor 2 results in an uncertainty in C of about 20%. For cross-sections which have a greater width than depth, C is expected to assume values between 0.4 and 1.0. For glaciers with some field data, the correction factor C can be calibrated. However, for glaciers with no a priori information, an estimation is necessary. Since C depends on the shape of the cross-section, one approach to estimate C is to consider glaciers for which similar bedrock shapes are expected. For Unteraargletscher, C was set to the mean value calibrated for similar valley glaciers, i.e., Rhonegletscher and Glacier de Zinal.

The sensitivity of the mean ice thickness with respect to changes in the gradients of the apparent mass balance and the calculated apparent ELA $\tilde{z}_0$ (Fig. 2.13b) is glacier-dependent. A change in $\tilde{z}_0$ has stronger effects for highly branched glaciers (e.g. Glacier de Zinal, Fig. 2.1a) than for glaciers with compact geometry (e.g. Rhonegletscher, Fig. 2.1b). This is due to the different distribution of area and volume, with branched glaciers having a larger area-to-volume ratio than glaciers with compact geometry. The difference between the mass balance gradients for the ablation and the accumulation area leads to a different sensitivity of the mean ice thickness for an increase or a decrease in the apparent ELA. The sensitivity of $\tilde{z}_0$ with respect to the gradients of the apparent mass balance is almost linear. The method is less sensitive to changes in the mass-balance gradients compared to the parameters of Glen's flow law. For the considered glaciers, identical gradients were assumed. This may not be realistic, but is a practical approach that diminishes the degrees of freedom. For an application of the method to glaciers in different climatic regions (e.g. maritime glaciers with high mass turnover or continental glaciers) additional input data (e.g. mass-balance measurements and rates of ice-thickness change) to determine the model parameters may be required. The comparison of calculated and measured ice thicknesses is the most direct way to validate the result. The validation with other types of data, such as flow-speed or ice volume flux measurements, is more complicated, since additional assumptions are required. The comparison between calculated and measured ice-flow velocities requires assumptions about unknown processes such as basal motion. Therefore, it provides only a plausibility check of the calculated glacier bedrock while indicating whether

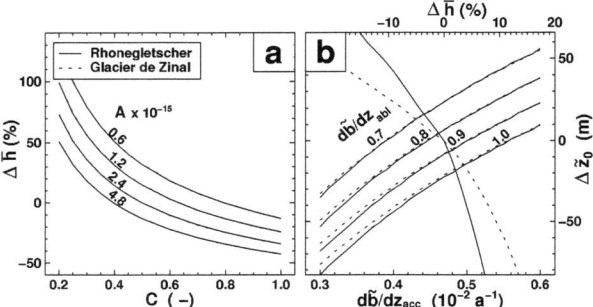

Figure 2.13: Sensitivity of calculated mean ice thickness $\bar{h}$ with respect to (a) correction factor C and flow rate factor A and (b) calculated apparent ELA $\tilde{z}_0$ (lines from top left to bottom right). The lines for Rhonegletscher and Glacier de Zinal are overlapping in (a). In (b), lines from bottom left to top right show the sensitivity of $\tilde{z}_0$ with respect to the gradients of the apparent mass balance $d\tilde{b}/dz_{\mathrm{acc/abl}}$.

the observations can be reproduced using realistic assumptions, rather than supplying direct information on the calculated bedrock.

In the case of Unteraargletscher, the comparison reveals an unrealistic bedrock for Profile 4 (Fig. 2.9), since no match between modelled and observed velocities can be achieved using realistic assumptions on basal sliding. The flow speed for Profile 4 is overestimated at the cross-sectional margins (Fig. 2.10b) and indicate an overestimation of the calculated ice thickness in this sector. This is confirmed by the comparison with the bedrock determined by radio-echo soundings.

The comparison of calculated and observed ice volume fluxes (Fig. 2.10a) is also influenced by assumptions about the basal sliding. According to Helbing (2005), Huss et al. (2007) assumed that basal motion accounts for 50% of the surface speed. The observed ice volume fluxes are systematically overestimated by our method (Fig. 2.10a). The overestimation for Profile 6 (located furthest from the glacier tongue), indicates an overestimation of $\tilde{b}$ in the accumulation zone. Huss et al. (2007) showed that, at present, only about 40% of the ablational mass loss of the tongue is compensated by the ice flux. The rate of ice-thickness change $\frac{\partial h}{\partial t}$ is therefore relatively high and the apparent mass balance $\tilde{b}$ differs significantly form the actual surface mass balance b. This means that the overestimation of the ice volume flux contributes to the correct reproduction of the ice thickness in the tongue region, which would, otherwise, be underestimated.

2.6 Conclusions

A method to estimate the ice-thickness distribution and the total ice volume of alpine glaciers from surface topography was presented. The accuracy inferred from point-to-point comparison of calculated and measured ice thickness is about 25%. Individual cross section areas can be reproduced within 20%.

The method is robust with respect to the chosen input geometry, and relatively insensitive with respect to the parameters describing the apparent glacier mass balance. The sensitivity of

2.6. CONCLUSIONS

the method is higher for the flow rate factor A and the correction factor C. C can be calibrated when measurements of the ice thickness or the ice-flow velocities are available; otherwise it has to be estimated. An estimate of C can be derived from the glacier type and geometry, since C is influenced by the bedrock shape. At present, the data base of ice-thickness measurements available for estimating C is still small and need to be expanded.

The analysis performed on four Swiss glaciers shows that the method is well suited to estimate the bedrock topography for glaciers where no direct measurements are available. The application of the method to different glaciers may require additional input data in order to correctly adjust the parameters describing the glacier mass turnover. The method is well suited to estimate the total ice volume of individual glaciers or small mountain ranges. The potential for analysis on a larger scale (e.g. questions related to sea level rise) is limited by the required (manual) digitization effort (glacier boundaries, flow lines and ice-flow catchments). For such analysis, the scaling approach may remain the only viable method (Raper and Braithwaite, 2005). In this regard, the presented method can be very valuable to enlarge the data basis that such scaling relations are based on.

Acknowledgments

Financial support for this study was provided by the BigLink project by the Competence Center Environment and Sustainability (CCES) of the ETH Domain. Swisstopo provided topographic maps and DEMs. H. Boesch evaluated the DEMs from aerial photographs. We thank all members of the field campaigns on which the radio-echo soundings were collected. P. Helfenstein and C. Zahno are acknowledged for evaluating the data from the field campaigns on Silvretta- and Rhonegletscher, respectively. S. Braun-Clarke edited the English. The constructive comments of the scientific editor H. Rott and the reviews of S. Raper and two anonymous reviewers helped to improve the manuscript.

Chapter 3

An estimate of the glacier ice volume in the Swiss Alps

Citation: Farinotti D., M. Huss, A. Bauder and M. Funk (2009). An estimate of the glacier ice volume in the Swiss Alps. *Global and Planetary Change*, 68, 225–231.

ABSTRACT: Changes in glacier volume are important for questions linked to sea-level rise, water resource management and tourism industry. With the ongoing climate warming, the retreat of mountain glaciers is a major concern. Predictions of glacier changes, necessarily need the present ice volume as initial condition, and for transient modelling, the ice thickness distribution has to be known. In this paper, a method based on mass conservation and principles of ice flow dynamics is applied to 62 glaciers located in the Swiss Alps for estimating their ice thickness distribution. All available direct ice thickness measurements are integrated. The ice volumes are referenced to the year 1999 by means of a mass balance time series. The results are used to calibrate a volume-area scaling relation, and the coefficients obtained show good agreement with values reported in the literature. We estimate the total ice volume present in the Swiss Alps in the year 1999 to be $74 \pm 9 \, km^3$. About 12 % of this volume was lost between 1999 and 2008, whereas the extraordinarily warm summer 2003 caused a volume loss of about 3.5 %.

3.1 Introduction

Glaciers are characteristic features of mountain environments and play an important role in various aspects. They are a key element of the water cycle of alpine catchments as they store water as snow and ice on many different time scales (Jansson et al., 2003), have a large contribution to the current rate of sea-level rise (e.g. Arendt et al., 2002; Raper and Braithwaite, 2006), and epitomize the "untouched environment", thus being precious for tourism industry. Warming of the climate system is unequivocal (Solomon and others, 2007) and for the Swiss Alps, a further temperature increase of $1.8°$ C in winter and $2.7°$ C in summer has been projected until the year 2050 (Frei, 2007). This causes major concern about the (partial) disappearance of alpine glaciers (e.g. Zemp et al., 2006; Huss et al., 2008b).

When assessing future glacier retreat, the current ice volume is the most important initial condition. In general, this information is available for a small number of glaciers only and is linked with major uncertainties. For transient glacier modelling, supplementary information about the ice thickness distribution is required. Obtaining this information is difficult, as direct measurement techniques, such as radio-echo soundings or borehole measurements, are laborious, necessarily restricted to a limited area, and the spatial inter- and extrapolation of the field data may lead to large uncertainties. For glaciers without direct ice thickness measurements, the total ice volume is often estimated with empirical relations between glacier area and volume (e.g. Erasov, 1968; Bahr et al., 1997) or glacier area and mean ice thickness (e.g Müller et al., 1976).

The glacier retreat in the Swiss Alps since the end of the Little Ice Age (around 1850) is well documented (e.g. Müller et al., 1976; Maisch et al., 2000; Glaciological Reports, 2008), and several studies provide projections for its future evolution (e.g. Schneeberger et al., 2003; Huss et al., 2007). Estimates of the total ice volume in the Swiss Alps, however, are scarce and based on the Swiss Glacier Inventory 1973 (SGI1973) by Müller et al. (1976) exclusively. Using two different empirical relations between glacier area and mean ice thickness, Müller et al. (1976) and Maisch et al. (2000) estimated the total ice volume of the Swiss Alps in 1973 to be $67\,km^3$ and $74\,km^3$, respectively.

A few more estimates exist for the total ice volume of the entire (not only the Swiss) Alps. Haeberli and Hoelzle (1995) applied a parametrization scheme based on vertical glacier extent, glacier length and glacier area to estimate the total ice volume around 1970 to be about $130\,km^3$. Putting this value into context with the Swiss ice volume is not straightforward, since the size distribution of glaciers in the European Alps is different from that one in the Swiss Alps.

Recently, Farinotti et al. (2009b) proposed a method based on mass conservation and principles of ice flow dynamics to estimate the ice thickness distribution of alpine glaciers from surface topography. In the current paper, we apply this method (here referred to as *ITEM* for *Ice Thickness Estimation Method*) and present an updated estimate of the total glacier ice volume of the Swiss Alps referenced to the year 1999. All available direct ice thickness measurements are integrated, none of them available for any of the previously mentioned studies addressing the ice volume in the Alps. Particular attention is focussed on assessing the uncertainty in the results. The volume estimation is carried out by applying (1) *ITEM* to all glaciers of the Swiss Alps with a surface area larger than $3\,km^2$ (59 glaciers in 1999) and some smaller selected glaciers for which ice thickness measurements are available (3 glaciers), and (2) an empirical volume-area scaling relation as proposed by Bahr et al. (1997) to glaciers smaller than $3\,km^2$. The sample of glaciers analyzed using *ITEM* covers 67 % of the total glacierized area. For these glaciers a complete glacier bedrock topography is generated on a 25 m grid, which, together with a surface elevation model, provides the ice thickness distribution.

3.2 Data

The most recent glacier inventory for Switzerland is the Swiss Glacier Inventory 2000 (SGI2000) by Paul (2004). SGI2000 was created from multi-spectral satellite imagery using geographic information system technology in combination with digital elevation models (DEMs). The data refer to the years 1998-1999 and are available in the *GLIMS Glacier Database* (www.glims.org). The detection of the glaciers contained in SGI2000 was performed with a semi-automatic procedure. This affects the accuracy of the glacier outlines in some cases (Paul, pers. comm., 2008). In order to verify the reliability of the SGI2000-data, a cross check with SGI1973 was performed, deciding for each individual glacier on the plausibility of the outline. Glaciers with evident discrepancies to SGI1973 or obviously missing in SGI2000 were manually re-digitized from the most recent topographic maps of the Swiss Federal Office of Topography (Swisstopo) with a scale of 1:25000 (Landeskarten 1:25 000). The re-digitization procedure was necessary for 40 glaciers. Since the definition of 'glacier' becomes somewhat loose for very small ice fields, glacierized areas smaller than 0.001 km^2 were not considered. With the revised inventory, the total glacierized area in Switzerland in 1999 is estimated to be 1063 ± 10 km^2.

High resolution DEMs obtained from airborne imagery are available for 18 glaciers (Bauder et al., 2007). Outlines for these glaciers were digitized from the aerial photographs. For additional 44 glaciers DEMs are taken from the Swisstopo product *DHM25*. For consistency with this data source, outlines are digitized from the corresponding topographic maps of Swisstopo with a scale of 1:25 000. All DEMs have a grid size of 25 m. Radio-echo soundings of the ice thickness are available for 11 glaciers. The data were acquired using the methods described in Bauder et al. (2003). In individual cases, additional information from earlier studies about the ice thickness was available. The data sets used are summarized in Table 3.1. In order to reference the ice volumes calculated by using *ITEM* to the year 1999, a mass balance time series was generated for 30 glaciers for the period 1990-2008. The series was obtained by applying the methods described in Huss et al. (2008a).

Table 3.1: Glacier surface area A and ice volume V for the considered glaciers. V_{year} is calculated by using *ITEM* and refers to the indicated year. V_{1999} is the ice volume referenced to 1999. *Data* indicates the data source for DEMs and outlines: 1) VAW-ETHZ, and 2) Swisstopo. When direct ice thickness measurements are available, the number of radio-echo sounding profiles (rp) is indicated. Radio-echo sounding profiles of Vadret da Morteratsch were provided by P. Huybrechts and O. Eisen. Glaciers are listed by decreasing V_{1999}.

#	Code	Glacier	Year	A_{year} km^2	V_{year} km^3	V_{1999} km^3	Data
1.	ALE	Grosser Aletschgletscher	1999	82.17	15.36 ± 4.52	15.36 ± 4.52	1) 13rp
2.	GOR	Gornergletscher	2003	55.23	5.85 ± 1.53	6.14 ± 1.97	1) 16rp
3.	UAA	Unteraargletscher	2003	22.71	3.75 ± 0.87	3.84 ± 0.88	1) 48rp
						Continued on next page	

Table 3.1 – Continued from previous page

#	Code	Glacier	Year	A_{year} km^2	V_{year} km^3	V_{1999} km^3	Data
4.	FIE	Fieschergletscher	1993	37.01	3.84 ± 0.96	3.70 ± 0.97	2)
5.	RHO	Rhonegletscher	2007	15.94	2.11 ± 0.38	2.23 ± 0.41	1) 20rp
6.	OAL	Oberaletschgletscher	1993	23.66	2.21 ± 0.55	2.05 ± 0.56	2)
7.	FIN	Findelgletscher	1995	18.93	1.89 ± 0.47	1.84 ± 0.48	2)
8.	ZMT	Zmuttgletscher	1995	20.24	1.57 ± 0.39	1.52 ± 0.40	2)
9.	UGR	Unterer Grindelwaldgletscher	2004	19.56	1.45 ± 0.37	1.50 ± 0.38	1) 5rp
10.	CRB	Glacier de Corbassière	2003	15.99	1.48 ± 0.65	1.48 ± 0.66	1) 11rp
11.	KND	Kanderfirn	1993	14.65	1.44 ± 0.36	1.39 ± 0.36	2)
12.	OTM	Glacier d'Otemma	1995	16.46	1.41 ± 0.35	1.37 ± 0.36	2)
13.	MRT	Vadret da Morteratsch	1991	16.58	1.25 ± 0.32	1.15 ± 0.32	2) 26rp
14.	TRF	Triftgletscher	2003	15.73	1.06 ± 0.26	1.11 ± 0.28	1)
15.	HUF	Hüfifirn	1997	13.45	1.12 ± 0.28	1.10 ± 0.28	2)
16.	ZIN	Glacier de Zinal	2006	13.41	0.89 ± 0.15	1.00 ± 0.24	1) 12rp
17.	GAU	Gauligletscher	1993	16.79	1.05 ± 0.26	0.99 ± 0.26	2)
18.	FRP	Glacier de Ferpècle	1995	13.22	0.97 ± 0.24	0.94 ± 0.24	2)
19.	ALL	Allalingletscher	2004	9.93	0.91 ± 0.23	0.90 ± 0.23	1)
20.	MMN	Glacier du Mont Miné	1995	11.28	0.89 ± 0.22	0.86 ± 0.22	2)
21.	FEE	Feegletscher	1995	16.27	0.88 ± 0.22	0.84 ± 0.22	2)
22.	TRT	Turtmanngletscher	1995	13.35	0.86 ± 0.21	0.83 ± 0.22	2)
23.	FOR	Vadret del Forno	1991	8.27	0.66 ± 0.16	0.61 ± 0.17	2)
24.	LNG	Langgletscher	1993	10.03	0.64 ± 0.16	0.61 ± 0.16	2)
25.	BRN	Glacier du Brenay	1995	8.96	0.62 ± 0.15	0.59 ± 0.15	2)
26.	PLM	Glacier de la Plaine Morte	1992	8.72	0.57 ± 0.14	0.53 ± 0.14	2)
27.	STN	Steingletscher	1993	8.74	0.56 ± 0.14	0.53 ± 0.14	2)
28.	GIE	Glacier du Giétro	2003	5.55	0.49 ± 0.08	0.50 ± 0.09	1) 5rp
29.	OGR	Oberer Grindelwaldgletscher	1993	10.11	0.49 ± 0.12	0.46 ± 0.12	2)
30.	MDR	Glacier du Mont Durand	1995	7.62	0.47 ± 0.12	0.45 ± 0.12	2)
31.	MAL	Mittelaletschgletscher	1993	8.42	0.48 ± 0.12	0.45 ± 0.12	2)
32.	TRN	Glacier du Trient	2005	5.87	0.41 ± 0.10	0.43 ± 0.10	1)
33.	RIE	Riedgletscher	1995	7.89	0.45 ± 0.11	0.43 ± 0.11	2)
34.	SCW	Schwarzberggletscher	2004	5.33	0.40 ± 0.10	0.42 ± 0.10	1)
35.	SAL	Glacier de Saleina	1995	8.82	0.43 ± 0.11	0.41 ± 0.11	2)
36.	GRS	Griesgletscher	2007	4.97	0.34 ± 0.11	0.40 ± 0.12	1) 14rp
37.	MOM	Glacier de Moming	2006	5.59	0.35 ± 0.09	0.39 ± 0.10	1)
38.	RSL	Rosenlauigletscher	1993	5.91	0.41 ± 0.10	0.39 ± 0.10	2)
39.	OAA	Oberaargletscher	1993	6.16	0.37 ± 0.09	0.34 ± 0.09	2)
40.	MCL	Glacier du Mont Collon	1995	6.58	0.35 ± 0.09	0.34 ± 0.09	2)
41.	MRY	Glacier de Moiry	1995	6.10	0.33 ± 0.08	0.32 ± 0.08	2)
42.	RSG	Vadret da Roseg	1991	8.79	0.36 ± 0.09	0.31 ± 0.09	2)
43.	CLA	Claridenfirn	1997	5.13	0.31 ± 0.08	0.31 ± 0.08	2)
44.	TGL	Tschingelfirn	1993	6.04	0.32 ± 0.08	0.30 ± 0.08	2)
45.	BAL	Baltschiedergletscher	1993	7.42	0.32 ± 0.08	0.29 ± 0.08	2)
46.	TSC	Vadret da Tschierva	1991	7.10	0.31 ± 0.08	0.27 ± 0.08	2)
47.	ARL	Haut Glacier d'Arolla	1995	5.18	0.26 ± 0.07	0.25 ± 0.07	2)
48.	CHE	Glacier de Cheillon	1995	4.07	0.24 ± 0.06	0.23 ± 0.06	2)

Continued on next page

3.3. METHODS 27

Table 3.1 – Continued from previous page

#	Code	Glacier	Year	A_{year} km^2	V_{year} km^3	V_{1999} km^3	Data
49.	DMM	Dammagletscher	2007	4.60	0.19 ± 0.05	0.23 ± 0.05	1)
50.	HLC	Hohlichtgletscher	1995	5.86	0.24 ± 0.06	0.22 ± 0.06	2)
51.	PLU	Vadret da Palü	1991	5.99	0.26 ± 0.06	0.22 ± 0.06	2)
52.	TFL	Glacier de Tsanfleuron	1998	3.30	0.21 ± 0.05	0.21 ± 0.05	2)
53.	WEI	Glacier du Weisshorn	2006	3.10	0.17 ± 0.04	0.19 ± 0.05	1)
54.	RTL	Rottalgletscher	1993	4.39	0.21 ± 0.05	0.19 ± 0.05	2)
55.	HBG	Hohbärggletscher	1995	3.38	0.19 ± 0.05	0.18 ± 0.05	2)
56.	SLV	Silvrettagletscher	2007	2.81	0.16 ± 0.03	0.18 ± 0.04	1) 9p
57.	BIS	Bisgletscher	1995	4.46	0.19 ± 0.05	0.18 ± 0.05	2)
58.	FRG	Furgggletscher	1995	3.98	0.17 ± 0.04	0.16 ± 0.04	2)
59.	WBL	Wildstrubelgletscher	1992	3.15	0.15 ± 0.04	0.13 ± 0.04	2)
60.	PRD	Paradiesgletscher	1995	2.40	0.07 ± 0.02	0.07 ± 0.02	2)
61.	ZPP	Zapportgletscher	1995	3.05	0.06 ± 0.01	0.05 ± 0.01	2)
62.	BAS	Ghiacciaio del Basodino	2002	2.84	0.03 ± 0.01	0.04 ± 0.01	1) 5rp
	TOTAL	(glaciers analyzed with ITEM)		**724.03**	$\mathbf{65.48 \pm 7.32}$	$\mathbf{64.92 \pm 7.58}$	

3.3 Methods

3.3.1 Ice thickness distribution and ice volume for glaciers $\geq 3\,\text{km}^2$

For glaciers with a surface area larger than $3\,\text{km}^2$ (Fig. 3.1), we estimate the ice thickness distribution and the ice volume using *ITEM*. The method is based on the principle of mass conservation: The mass balance distribution of a glacier is balanced by the ice flux divergence and the resulting surface elevation change. In general, the distributions of mass balance and surface elevation change are unknown. Therefore, a variable called "apparent mass balance" $\tilde{b}$, defined at any point as the difference between the glacier surface mass balance and the rate of ice thickness change, is introduced (Farinotti et al., 2009b). The definition allows fulfilling the mass conservation equation integrated over the glacier (e.g. Paterson, 1994) without the requirement of a steady-state assumption. $\tilde{b}$ is assumed to vary linearly with altitude according to two distinct gradients for the ablation and the accumulation area. The distribution of $\tilde{b}$ is used to calculate a cumulative ice volume flux for defined ice flow lines which need to be digitized manually. The ice volume flux is converted into an ice thickness using an integrated form of Glen's (1955) flow law and the shallow ice approximation (e.g. Hutter, 1983). The resulting ice thickness is spatially interpolated over the glacier using the glacier outlines as boundary condition with zero ice thickness. Finally, the ice thickness distribution is adjusted with a factor that accounts for the local surface slope. For details refer to Farinotti et al. (2009b).

Five parameters need to be adjusted: (1, 2) Two vertical gradients of the apparent mass balance, $d\tilde{b}/dz_{acc}$ for the accumulation area and $d\tilde{b}/dz_{abl}$ for the ablation area, defining the distribution of $\tilde{b}$, (3) a debris coverage reduction factor f_{debris} which accounts for the influence of debris cover on mass balance, (4) the flow rate factor A of Glen's flow law, and (5) a correction factor C which accounts for (i) the approximation of the shear stress distribution by a linear relation (Nye, 1965), (ii) the approximation of the specific ice volume flux at the centre of the profile with the mean ice volume flux across the profile, (iii) the influence of basal sliding,

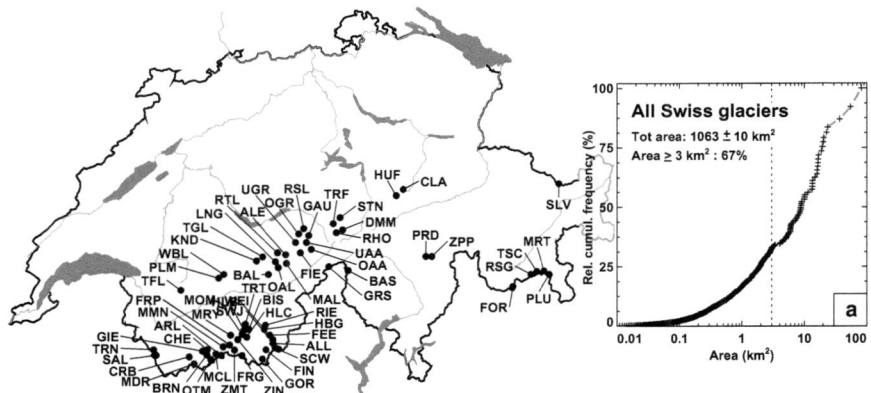

Figure 3.1: Glaciers for which the ice volume was calculated by using *ITEM*. Glaciers are labeled with the same code as in Table 3.1. (a) Relative cumulative frequency of glacier surface area of all glaciers in the Swiss Alps.

assuming a linear relation between basal and deformation velocity (e.g. Gudmundsson, 1999), and (iv) the uncertainties in the flow rate factor A.

Except for the correction factor C, we chose the same parameters for all glaciers (Table 3.2). The vertical gradients $d\widetilde{b}/dz_{\text{acc/abl}}$ were determined by analyzing the altitudinal distribution of mass balance and ice thickness change of 12 glaciers. We chose glaciers that are not debris covered, have an elevation range larger than 500 m and have two available DEMs in a time span of at least 20 years. The altitudinal distribution of the surface mass balance is calculated by using the methods described in Huss et al. (2008a). The distribution of the rate of ice thickness change is calculated by differencing the two successive surface DEMs. The aggregation of the data in 10 m surface elevation bands is then used to determine two vertical gradients for $\widetilde{b}$, one for the accumulation and one for the ablation zone, by linear regression. We found typical gradients of $-4.0 \cdot 10^{-3}\, \text{a}^{-1}$ for the ablation and $-2.5 \cdot 10^{-3}\, \text{a}^{-1}$ for the accumulation area. f_{debris} was set equal to 0.5 (Huss et al., 2007), A was chosen to be $2.4 \cdot 10^{-15}\, \text{kPa}^{-3}\, \text{s}^{-1}$ (Hubbard et al., 1998; Gudmundsson, 1999). Two cases are distinguished for the correction factor C: (1) For glaciers with available radio-echo sounding profiles (cf. Table 3.1), C is calibrated to obtain the best possible agreement between calculated and measured ice thickness. This is done individually for each profile. Between two profiles, C is interpolated linearly with distance. (2) For glaciers with no ice thickness measurements, $C = 0.53$ was obtained from the mean value calibrated for all glaciers with ice thickness measurements.

The relative uncertainty in the ice thickness distribution calculated by using *ITEM* is expressed in terms of σ_h/h, where σ_h is the standard deviation of the ice thickness h. Assuming normal-distributed errors, the 95 % confidence interval of the ice thickness h_i calculated at any location i, is then $h_i \pm 2\frac{\sigma_h}{h} h_i$.

Depending on the availability of direct ice thickness measurements, we estimate σ_h/h by distinguishing three cases:

1) For glaciers with evenly distributed radio-echo sounding profiles over the surface area (9 glaciers), σ_h/h is estimated as follows: When the glacier has n available radio-echo

3.3. METHODS

Table 3.2: Parameter values and units for *ITEM*.

Parameter	Symbol	Value	Unit
$d\widetilde{b}/dz$ accumulation zone	$\frac{db}{dz}\big|_{acc}$	$-2.5 \cdot 10^{-3}$	a^{-1}
$d\widetilde{b}/dz$ ablation zone	$\frac{db}{dz}\big|_{abl}$	$-4.0 \cdot 10^{-3}$	a^{-1}
Debris reduction factor	f_{debris}	0.5	–
Flow rate factor	A	$2.4 \cdot 10^{-15}$	$kPa^{-3} s^{-1}$
Correction factor	C	$0.53^{a)}$	–

a) For glaciers without direct measurements. Else calibrated.

sounding profiles, n glacier bedrock topographies are calculated by using $n-1$ profiles to calibrate the correction factor C. At each step, the measured, h_{meas}, and calculated, h_{calc}, ice thicknesses along the profile not used for calibration are stored. σ_h/h is then approximated with the normalized root mean square error of all m stored h_{meas} and h_{calc}:

$$\frac{\sigma_h}{h} \approx \sqrt{\frac{1}{m} \sum_{i=0}^{m} \left(\frac{h_{meas,i} - h_{calc,i}}{h_{meas,i}} \right)^2}. \tag{3.1}$$

We found $\sigma_h/h = 15\%$.

2) For glaciers without direct measurements, σ_h/h is estimated by calculating a bedrock topography for glaciers where measurements are available and setting $C = 0.53$. Comparing measured and calculated ice thicknesses, an uncertainty is computed according to Equation 3.1. This uncertainty is then assumed to be transferable to glaciers with no measurements. We found $\sigma_h/h = 35\%$.

3) For glaciers where the available radio-echo sounding profiles are not evenly distributed over the glacier, but clustered in a specific region (i.e. Glacier de Zinal, Unterer Grindelwaldgletscher and Gornergletscher), we found that the information contained in the available radio-echo sounding profiles does not improve the accuracy of the ice thickness distribution calculated for the glacier section without measurements. Thus, the uncertainty in the results is estimated by combining the approaches (1) and (2) by subdividing the glacier in two sub-regions: One, where the radar profiles are clustered and for which the uncertainty is estimated using (1), and one with no radar profiles and for which the uncertainty is estimated using (2).

The accuracy in the digitized glacier areas is high. Based on multiple digitization of glacier outlines, we estimate σ_A/A to be equal to 3 %. The total uncertainty in the ice volume V calculated by using *ITEM* for a given glacier is then

$$\frac{\sigma_{V,ITEM}}{V_{ITEM}} = \sqrt{\left(\frac{\sigma_h}{h}\right)^2 + \left(\frac{\sigma_A}{A}\right)^2}. \tag{3.2}$$

For the 62 glaciers analyzed by using *ITEM*, we found $\sigma_{ITEM}/V_{ITEM} = 27\%$.

For each glacier, the bedrock topography calculated by using *ITEM* is based on the surface of the latest DEM available, corresponding to the minimal glacier extent (Table 3.1). Thus, the calculated ice volumes refer to different years. To obtain a data set which consistently refers to the reference year 1999 (given by the date of SGI2000), we adjust the calculated ice volumes

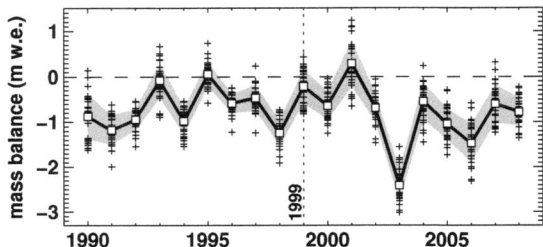

Figure 3.2: Mass balance time series $\bar{b}$ used for referencing the ice volumes calculated by using *ITEM* to the year 1999. The solid line represents the 30-glacier average with one standard deviation (gray band). Values for individual glaciers are shown with crosses.

by using a time series of annual mass balance (Fig. 3.2). For a glacier with a calculated ice volume V_y for a given year y and a mean specific mass balance $\bar{b}_i$ in the year i, we calculate the ice volume V_{1999} for the reference year as:

$$V_{1999} = \begin{cases} V_y + \sum_{i=y}^{1999} A \cdot \bar{b}_i & \text{if } y < 1999 \\ V_y - \sum_{i=1999}^{y} A \cdot \bar{b}_i & \text{if } y > 1999 \end{cases} \quad (3.3)$$

where A is the glacier area, which is kept constant for the time interval y to 1999. The time series of $\bar{b}_i$ includes 21 glaciers analyzed using *ITEM*. For the remaining glaciers, an averaged mass balance time series is used, calculated as the arithmetic mean of all data in each year. The additional uncertainty in the ice volume $\sigma_{V,\text{shift}}$ induced by referencing it to the year 1999 is calculated from the uncertainty σ_{A_c} due to assumption of a constant area in the interval y to 1999 and the uncertainty $\sigma_{\bar{b}}$ in the mass balance time series. Based on the average interval over which an ice volume calculated by using *ITEM* is shifted and the average glacier size, we estimate σ_{A_c}/A to be less than 3 %. Note that σ_{A_c} has no direct relation to the uncertainty σ_A in the digitized glacier areas, although the two quantities have the same numeric value. $\sigma_{\bar{b}_n}$ is assumed to be 0.2 m w.e. a^{-1} for glaciers with mass balance records (Dyurgerov and Meier, 2002; Huss et al., 2009a) and is estimated as the average standard deviation of the mass balance of each year (0.35 m w.e. a^{-1}) for glaciers where the averaged series is applied. Based on this considerations, $\sigma_{V,\text{shift}}$ is calculated as:

$$\sigma_{V,\text{shift}} = A \cdot \sqrt{\sum_{i=y}^{1999} \left(\left(\bar{b}_i \frac{\sigma_{A_c}}{A} \right)^2 + (\sigma_{\bar{b}})^2 \right)}. \quad (3.4)$$

Finally, the uncertainty in the ice volume V_{1999} calculated for the reference year 1999 is given by:

$$\sigma_{V,1999} = \sqrt{\sigma_{V,y}^2 + \sigma_{V,\text{shift}}^2}. \quad (3.5)$$

3.3.2 Ice volume for glaciers $< 3 \, \text{km}^2$

The application of *ITEM* requires (manual) digitization of glacier boundaries, ice flow lines and ice flow line catchments. An application to all glaciers inventoried in SGI2000 is, thus, impracticable. The 62 glaciers analyzed by using *ITEM* cover 67 % of the total glacier area. For the remaining 33 %, an alternative approach has to be chosen. Based on a scaling analysis of the conservation equations for mass and momentum, Bahr et al. (1997) showed the physical basis of volume-area scaling relations for estimating ice volumes of alpine glaciers. Such relations are power laws of the form

$$V = c \cdot A^\gamma, \tag{3.6}$$

where V (km^3) is the total ice volume of a glacier with surface area A (km^2), γ and c empirical constants. To estimate the ice volume of the glaciers not analyzed using *ITEM*, we use this approach. We consider each of the analyzed glaciers as one single surface; individual glacier branches are not split. The parameters c and γ are determined by linear regression of the logarithmic area-volume data set obtained for the 62 glaciers analyzed using *ITEM*. We found $c = 0.025$ and $\gamma = 1.41$ (Fig. 3.3).

The 95 % confidence interval of the ice volume V_{Bahr} predicted for an individual glacier of area A using Equation 3.6 is given by the prediction interval:

$$V_{\text{Bahr}} \cdot \exp\left(\pm q \sqrt{\sigma_r^2 + \varepsilon_s^2}\right), \tag{3.7}$$

where q is the 0.95-quantile for a t-distribution with $n - 2 = 60$ degrees of freedom (q=2.000), σ_r^2 is the variance of the n residuals $r_i = V_{\text{ITEM},i} - V_{\text{Bahr},i}$ and

$$\varepsilon_s = \sigma_r \sqrt{\frac{1}{n} + \frac{(A - \overline{A})^2}{\sum_{i=0}^{n}(A_i - \overline{A})^2}}, \tag{3.8}$$

where $\overline{A}$ is the average of the n glacier areas A_i used for the calibration of the relation and A the area of the glacier for which a volume is predicted. Assuming $\sigma_{\text{Bahr}} = \frac{1}{2}q\sqrt{\sigma_r^2 + \varepsilon_s^2}$, one can write $\ln(V_{\text{Bahr}}) \pm 2\sigma_{\text{Bahr}}$ for the 95 % confidence interval. We found $\sigma_{\text{Bahr}}/V_{\text{Bahr}} = 12.7\%$. Note that the confidence interval refers to the logarithmic relation and is, thus, not symmetric around the ice volume V. On average, the confidence interval ranges from 75 % above to 43 % below the calculated value V (Fig. 3.3).

3.4 Results and discussion

Applying the methods described above, we estimate the total glacier ice volume present in the Swiss Alps in 1999 to be $74 \pm 9\,\text{km}^3$. This corresponds to a mean ice thickness of $70 \pm 8\,\text{m}$. About 88 % of the volume ($65 \pm 5\,\text{km}^3$) is stored in the 59 largest glaciers (glaciers with a surface area $A \geq 3\,\text{km}^2$, Table 3.3). The glacier sample analyzed with the empirical volume-area scaling relation (glaciers with $A < 3\,\text{km}^2$) contributes with 33 % to the total surface area (Fig. 3.1a) and contains 12 % of the total volume (Fig. 3.4). For 6 of the 10 largest glaciers, which contribute together to more than half of the total estimated ice volume, direct ice thickness measurements were available to determine the ice volume. This causes, together with the assumption of normally distributed uncertainties in the individual ice volumes, the confidence interval of the total ice volume to become relatively small (12 %), even though significantly larger uncertainties are assessed for individual glaciers.

For the 62 glaciers analyzed using *ITEM*, a glacier bedrock topography is generated on a 25 m grid. The resulting data set is unique in its coverage and resolution. As an example, we present the results for the glaciers in the hydrological drainage basin of the Massa

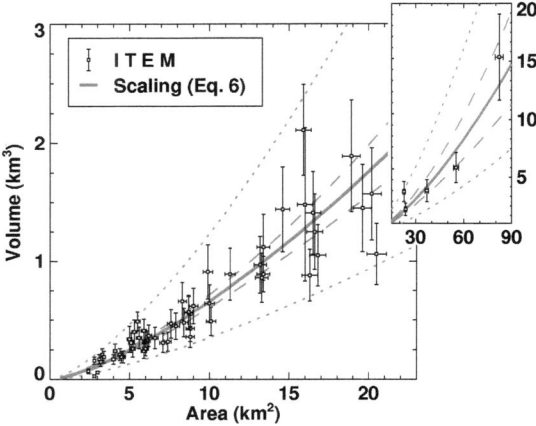

Figure 3.3: Relation between glacier area and estimated ice volume of the glaciers shown in Figure 3.1. The enlargement shows UAA, OAL, FIE, GOR and ALE (from left to right). For the volume-area scaling relation (line), the 95% confidence interval of the fitted curve (dashed) and the 95% prediction interval (dotted) are shown.

Table 3.3: Total area A_{tot} and volume V_{tot} referenced to 1999.

	Count	A_{tot} (km^2)	V_{tot} (km^3)
Glaciers with $A \geq 3$ km^2	59	715 ± 8	65 ± 8
Glaciers with $A < 3$ km^2	1424	348 ± 3	9 ± 4
TOTAL	1483	1063 ± 10	74 ± 9

river, which includes Grosser Aletschgletscher, the largest glacier of the European Alps (Fig. 3.5). The drainage basin covers 195 km^2 and comprises three main valley glaciers (Grosser Aletschgletscher, Mittelaletschgletscher and Oberaletschgletscher) and several smaller ice masses. The total glacierized surface area was 125 km^2 in 1999, which corresponds to 12 % of the ice-covered area of Switzerland. For Grosser Aletschgletscher, direct ice thickness measurements are available for 12 radio-echo sounding profiles. However, the information is generally constrained to the flanks of the main trough. In this configuration, *ITEM* is a powerful tool for integrating the available information which would be insufficient for using a simple interpolation scheme of the bedrock topography based on direct measurements. According to *ITEM*, the drainage basin has a total ice volume of 18 ± 7 km^3 which corresponds to 24 % of the ice volume in the Swiss Alps. This shows the importance of large glaciers in terms of ice volume. The largest ice thickness is at Konkordiaplatz, the zone where the three main branches composing Grosser Aletschgletscher merge. The ice thickness at this location was already investigated in earlier studies. Thyssen and Ahmad (1969) draw an ice thickness map based on seismic measurements pointing out a pronounced local overdeepening. The maximum ice thickness was estimated to be about 890 m, corresponding to a bedrock at 1850 m a.s.l. However, ice thicknesses exceeding 800 m were assessed in an area of about 600 x 600 m only. In 1990-1991, VAW-ETHZ drilled boreholes in the same area, three of them reaching bedrock (Hock et al.,

3.4. RESULTS AND DISCUSSION

1999). One of the boreholes reached a depth of 900 m, whereas the other two reached the bedrock at already about 750 m. The results confirmed that ice thicknesses larger than 800 m are attributable to a relatively small topographic feature. However, this type of features cannot be recognized by *ITEM* and thus, the maximum ice thickness calculated for the Konkordiaplatz (about 750 m) should not be interpreted as a failure of the method.

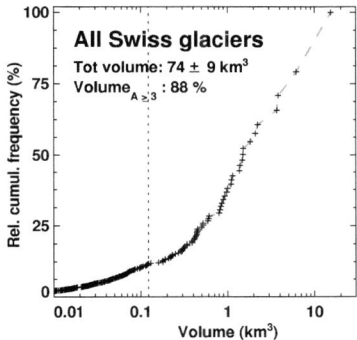

Figure 3.4: Relative cumulative frequency of calculated volume for all glaciers in the Swiss Alps. The dashed line indicates the volume corresponding to an area $A = 3\,\text{km}^2$ (Eq. 3.6).

The parameters determined by regression analysis for the volume-area scaling relation (Eq. 3.6), and in particular the power exponent $\gamma = 1.41$, are in good agreement with earlier published values (e.g Macheret et al., 1988; Zhurovlyev, 1985; Chen and Ohmura, 1990a). Although Bahr et al. (1997) showed that, for physical reasons, γ should assume a close range of values, the result is surprising, since the data set used is independent of those used in earlier studies. Analyzing a sample of synthetically generated glaciers, Radić et al. (2007) pointed out that for glaciers in non-steady-state conditions, γ may differ significantly from the theoretical value. Although in our analysis we considered glaciers out of steady-state, this could not be confirmed.

With the calibrated volume-area scaling relation, the volumes of the 62 glaciers analyzed using *ITEM* can be reproduced with a standard deviation of 37 % (Fig. 3.3). Due to the power relation between glacier area A and volume V, the prediction interval of the volume-area scaling relation is not symmetric. For individual glaciers, the 95 % confidence interval for the predicted ice volume ranges from 75 % above to 43 % below the predicted value (Fig. 3.3).

The uncertainty in the ice volumes calculated by using *ITEM* is mainly determined by the available data sets. The uncertainty in the ice volume calculated for a given year y, can be significantly reduced when direct measurements are available, whereas the uncertainty added by referencing the result to 1999 plays a minor role. Even on glaciers where an averaged mass balance time series is applied, $\sigma_{V,\text{shift}}$ is less than 3 % of the glacier volume.

The mass balance time series allows the total glacier ice volume calculated for the Swiss Alps to be put in context of climate change. Based on Equation 3.3, we estimate that in the time span 1999–2008, the total ice volume has decreased by about 12 %. Due to the assumption of a constant glacier area in this time interval, the value has to be considered as an upper boundary. The volume loss occurring during the particularly warm year 2003 ($\bar{b} = -2.40\,\text{m w.e.}$) is estimated as about 3.5 % of the total glacier ice volume of the Swiss Alps, corresponding to 2.6 km^3 of ice.

3.5 Conclusions

Applying *ITEM* to glaciers with a surface area larger than 3 km^2 (59 glaciers), integrating all available direct measurements of the ice thickness and calibrating a volume-area scaling relation, the total glacier ice volume present in the Swiss Alps by the year 1999 is estimated to be 74 ± 9 km^3. This corresponds to an average ice thickness of 70 ± 8 m. About 88 % of the total ice volume is stored in the 59 largest glaciers.

The accuracy of the total ice volume is affected by the uncertainty in (1) estimating the ice thickness distribution using *ITEM*, (2) determining the glacier area of individual glaciers, and (3) applying the volume-area scaling relation. The uncertainty induced by referencing the individual ice volumes to a given year and the uncertainty due to the incompleteness of the SGI2000 are of minor importance. The comparison of the ice volumes calculated by using *ITEM* and the volume-area scaling relation indicates that the scaling approach is well suited for estimating total volumes of glacier samples.

Applying an averaged mass balance time series, we estimated that about 12 % of the total ice volume present in the Swiss Alps by 1999 was lost during the period 1999 to 2008 and that the particularly warm summer 2003 caused a volume loss of about 3.5 %.

The strength of *ITEM* is that a complete bedrock topography is generated and that direct ice thickness measurements can easily be integrated. The glacier bedrock topographies obtained offer the opportunity for further investigations. The combination of *ITEM* and a volume-area scaling relation is a valuable approach to estimate the total glacier ice volume of mountain ranges.

Acknowledgments

Financial support for this study was provided by the BigLink project of the Competence Center Environment and Sustainability (CCES) of the ETH Domain. Swisstopo provided topographic maps and DEMs. H. Bösch evaluated the DEMs from aerial photographs. F. Paul provided valuable support in topics related to the SGI2000 data set. P. Huybrechts and O. Eisen kindly provided the radio-echo sounding profiles of Vadret da Morteratsch. We thank all members of the field campaigns on which the radio-echo soundings of VAW-ETHZ were collected, especially the Swiss Army Air Force for the logistic support. S. Braun-Clarke and K. Hutter edited the English. We acknowledge P. A. Pierazzoli for the work as editor and R. J. Braithwaite and H. Blatter for the reviews.

3.5. CONCLUSIONS

Figure 3.5: Ice thickness distribution calculated by using *ITEM* in the hydrological basin of the Massa river. Available radio-echo sounding profiles are shown. Hatched areas are analyzed using the scaling relation (Eq. 3.6).

Chapter 4

Snow accumulation distribution inferred from time-lapse photography and simple modelling

Citation: Farinotti D., J. Magnusson, M. Huss and A. Bauder (2010). Snow accumulation distribution inferred from time-lapse photography and simple modelling. *Hydrological Processes*, 24, 2087–2097.

ABSTRACT: The spatial and temporal distribution of snow accumulation is complex and significantly influences the hydrological characteristics of mountain catchments. Many snow redistribution processes, such as avalanching, slushflow or wind drift, are controlled by topography, but their modelling remains challenging. In situ measurements of snow accumulation are laborious and generally have a coarse spatial or temporal resolution. In this respect, time-lapse photography reveals itself as a powerful tool for collecting information at relatively low cost and without the need for direct field access. In this paper, the snow accumulation distribution of an Alpine catchment is inferred by adjusting a simple snow accumulation model combined with a temperature-index melt model in order to match the modelled melt-out pattern evolution to the pattern monitored during an ablation season through terrestrial oblique photography. The comparison of the resulting end-of-winter snow water equivalent distribution with direct measurements shows that the achieved accuracy is comparable to that obtained with an inverse-distance interpolation of the point measurements. On average over the ablation season, the observed melt-out pattern can be reproduced correctly in 93 % of the area visible from the fixed camera. The relations between inferred snow accumulation distribution and topographic variables indicate large scatter. However, a significant correlation with local slope is found and terrain curvature is detected as a factor limiting the maximal snow accumulation.

4.1 Introduction

When studying characteristics of alpine catchments, snow is an important component. It can strongly increase the albedo of a surface, reduce its roughness, insulate the underlying ground from the atmosphere and store or release large amounts of water (Essery et al., 1999). In alpine environments, snow cover dynamics often dominate water resource formation, storage and release (Lehning et al., 2006). Modelling the snowpack and its evolution is thus of interest for a wide range of applications, including flood and avalanche forecasting or simulations of effects caused by climate change. However, modelling solid precipitation in the Alps is challenging. The Alpine chain modifies the structure of overlying synoptic systems through cyclogenetic processes (e.g. Mesinger and Pierrehumbert, 1986), and the effect of complex patterns of mountain ridges and valleys modulates the distribution of precipitation via orographic upslope lifting and the development of local mesoscale circulations (Wallén, 1970). At smaller scales, the precipitation distribution is affected considerably by local topographic effects (e.g. Spreen, 1947; Basist et al., 1994). As a result, snow accumulation in mountainous terrain is complex.

The most important topographic factor controlling the amount of deposited snow is elevation (e.g. Seyfried and Wilcox, 1995). The strong altitudinal gradient is due to the relation between altitude and both, temperature and precipitation. Once on the ground, snow undergoes a variety of redistribution processes, such as wind drift (e.g. Kind, 1981; Dadic et al., 2010), avalanching and slushflow (e.g. Luckman, 1977). The resulting snow distribution patterns are often similar over time due to the topographic control of the redistribution processes (e.g. Elder et al., 1991; Luce et al., 1998).

Many approaches exist to model wind-drift processes (e.g. Schmidt, 1986; Liston and Sturm, 1998) some of them achieving a high grade of process-detail description, such as particle-bed collisions, particle motion and particle-wind feedback (e.g. Doorschot et al., 2001; Nemoto and Nishimura, 2004; Lehning et al., 2008). However, precipitation modelling is often only one component of a more complex model, for instance, one aiming at hydrological analysis (e.g. Schäfli et al., 2005; Lehning et al., 2006). In such cases, simpler and computationally cheaper approaches are appreciated.

Although information on the snow cover is often available at high temporal resolution from automatic stations at individual points (Egli, 2008), the spatial resolution of this information is generally coarse. In situ measurement of snow accumulation with high spatial resolution is very costly in terms of time, manpower, and expense. For this reason, considerable effort has focused on developing methods for estimating the distribution of snow properties from remotely sensed data (e.g. Dozier and Painter, 2004; Machguth et al., 2006). Microwave remote sensing has shown promise for evaluating different properties, including snow depth and snow density (Shi and Dozier, 2000a,b). A review of methods developed for measuring snow and glacier ice properties through satellite remote sensing was presented by König et al. (2001).

The idea of inferring snow accumulation distribution from the observed melt-out pattern has been presented before in different studies (e.g. Martinec and Rango, 1981; Davis et al., 1995; Turpin et al., 1999) but often the observed snow cover evolution was used for the validation of distributed melt models only (e.g Leavesley and Stannard, 1990; Blöschl et al., 1991; Mittaz et al., 2002). Moreover, in all mentioned studies the evolution of melt-out was observed either by satellite or aerial photography, which are relatively expensive sensors, not easy to operate, and generally with a low temporal resolution.

In comparison, conventional photography can be a powerful tool for collecting information in an easy way and at relative low cost (Corripio, 2004). If the information is referenced pre-

4.2. STUDY SITE AND DATA

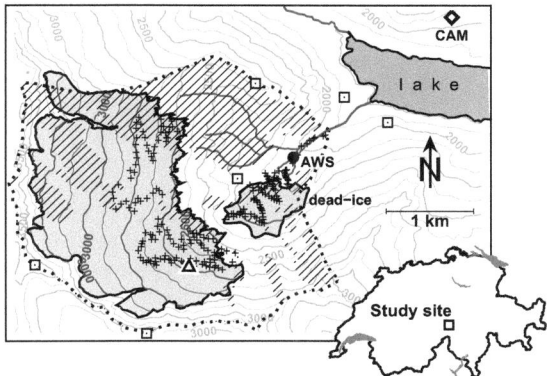

Figure 4.1: Study site overview. Terrain elevation is indicated by 100 m contour lines and the glacier is highlighted in light gray. The small glacier part at lower elevations is a dead-ice body. The positions of the automatic weather station (AWS) and the fixed camera (CAM) are shown. Regions of the catchment (dotted boundary) not seen from the camera are hatched. The locations of the snow depth measurements and the snow pit are shown by crosses and a triangle, respectively. Ground control points used for georeferencing are shown with a ⊡ symbol.

cisely in space, photography becomes a valid tool for quantitative analysis. Such applications are well developed in photogrammetry, but not in conventional photography.

In this paper we combine the pictures taken during a melt-out season by a fixed conventional photographic camera with a distributed melt and accumulation model, in order to infer the snow accumulation distribution in a small catchment in the Swiss Alps. The inferred distribution, which is assumed to be constant over time, is then used to build up the snow cover during one winter season. The resulting end-of-winter snow water equivalent (SWE) distribution is validated against direct snow depth measurements and the distribution of the parameter controlling the snow accumulation is correlated with topographic variables.

4.2 Study site and data

Our analyses are performed in the alpine catchment of Dammagletscher, in the central Swiss Alps (Fig. 4.1). The catchment had a glacierization of 50 % in 2007, is 9.1 km^2 in size and has an elevation range between 1940 and 3630 m a.s.l.. The hydrological regime of the study area is dominated by the presence of the glacier and the seasonal snow cover. The topography of the study site is characteristic for the complexity of mountainous terrain, with small flat surfaces having relatively high surface roughness (e.g. glacier forefield) alternated with steep and high side walls clearly confining the catchment (e.g. west ridge). In the forefield, a debris-covered dead-ice body with virtually no ice flow dynamics represents the most prominent feature. The catchment is the focus of a larger multidisciplinary project of the Competence Centre of Environment and Sustainability (CCES) at ETH Zurich with the aim of providing a detailed description of the processes occurring at the biosphere-hydrosphere-geosphere interface.

The topography of the study site is known from a high-accuracy digital elevation model

(DEM) obtained from airborne photogrammetry in 2007. For the glacier-covered part of the catchment, two additional DEMs are available for 1939 and 1959 from digitized topographic maps. Glacier outlines for the three DEMs have been determined from the respective data source.

An automatic weather station (AWS) installed in the glacier forefield at 2025 m a.s.l. (Fig. 4.1) recorded average air temperature and precipitation sums in 10 minutes intervals since October 2007. Additional temperature and precipitation data are available from the network of meteorological stations of the Federal Office of Meteorology and Climatology (MeteoSwiss).

Pictures of the study site were taken daily during the period between mid-May and the end of September 2008 with a conventional Olympus OM-2N photographic camera. The camera was located about 2 km from the catchment outlet and the field of view covered 62 % of the catchment area (Fig. 4.1). Pictures were taken automatically with a 28 mm unfiltered lens every day at 9 AM local time. The timing was chosen based on the incidence angle of the sunlight in order to get the best possible contrast in the region of interest. The camera was operated with Kodak Kodachrome ISO 64/19° 24×36 mm films, which had to be replaced manually about once a month.

On May 15, 2008 – just before the installation of the camera – a field survey was conducted to measure the SWE distribution in the catchment. A total of 227 snow depth measurements was collected and one snow pit was dug for determining snow density (Fig. 4.1). Density was determined by weighing snow samples collected with a metallic tube of known volume with a spring-scale balance. Snow samples were taken in 50 cm depth intervals and for each depth the procedure was repeated three times. The average density of the snowpack was then calculated as the arithmetic mean of all measurements.

4.3 Methods

In order to get quantitative information from the sequence of pictures taken using the fixed camera, a georeferencing of the images was necessary. The snow cover evolution was then modelled combining a temperature-index melt model and an accumulation model, both driven by daily temperature and precipitation time series. The accumulation model was adjusted in order to reproduce the short-term evolution of the snow cover, i.e., the melt-out pattern during a single season, while the melt model was adapted in order to match the glacier volume change observed during the period 1939–2007, reproducing, thus, the evolution on a longer time scale.

4.3.1 Image processing

For image processing, a method presented by Corripio (2004) was applied. The result is a snow cover map with the same resolution as the available DEMs, indicating for every grid cell whether the cell is snow-covered, snow-free or not visible from the position of the fixed camera. The image processing procedure consists of five steps:

A) Selection of images: In a first step, images suitable for further processing were selected. The selection had the objective to provide a homogeneous chronosequence of the melt-out pattern, not biased by summer snowfall events. Images taken during cloudy or foggy days as well as images with clouds obscuring the view of the catchment were discarded. The images were chosen in order to obtain the best possible contrast for snow detection. This resulted in the selection of 8 pictures, documenting the evolution of the melt-out

4.3. METHODS

during the period May 23 to July 30, 2008 in about 10 day-intervals. Pictures taken after July 30 were discarded because of important snowfalls interrupting the melt-out pattern evolution.

B) Relative image adjustment: When operating conventional photo cameras, small shifts in the camera orientation, e.g., after the replacement of the film, are common. In order to correct these small variations in orientation, images chosen for further processing were adjusted relative to each other (Fig. 4.2a). For this purpose, four reference points were manually selected in each of the 8 pictures. The pictures were then rotated and deskewed in order to co-register the four points.

C) Detection of snow-covered areas: Whereas the detection of snow is relatively easy on bare soils or dark surfaces, at least during cloud-free days, the distinction between snow and ice is more difficult in the visible band (e.g. Good and Martinec, 1987). Automatic snow detection procedures are, thus, generally unreliable. Since the camera used had no infrared sensor, the detection of snow-covered areas was performed manually. This was done by coloring zones interpreted as snow-covered using a commercial image-processing package (Fig. 4.2b). The manual snow detection required about 30 minutes per processed image.

D) Georeferencing of images: The georeferencing of the images was performed using a procedure developed by Corripio (2004). Basically, the georeferencing procedure requires the definition of a function relating two-dimensional pixels in the photograph to three-dimensional points in the DEM. This is achieved by applying a perspective projection of the image pixels after transforming the coordinates of the DEM to the camera coordinate system. In this way, a "virtual" photograph of the DEM is produced, that is, a two-dimensional representation of the relief information contained in the DEM, as seen from the point of view of the camera. By scaling this representation according to the resolution of the photograph, the necessary correspondence between pixels in the image, screen coordinates of the perspective projection of the DEM and their geographic location, can be established. The correct scaling functions are defined by matching the position of six ground control points (three manually installed ones and three well defined mountain tops for which the position is known from topographic surveys, Fig. 4.1 and 4.2a) to the corresponding locations on the images. The result is a georeferenced ortho-image of reflectance values. In case of visible photography, used in this study, reflectance refers to the three optical bands (red, green and blue) to which the film is sensitive (Fig. 4.2c). The same method is, however, applicable to images representing other wavelengths as well (e.g. infrared photography). For further details on the georeferencing procedure refer to Corripio (2004).

E) Compilation of snow cover maps: The final step consists of the compilation of a series of gridded maps with same coordinates and resolution as the DEM, containing the information about the snow cover of each individual grid cell. This is achieved by detecting the previously marked snow-covered areas through their reflectance value in the georeferenced images. Zones of the DEM which are not seen from the position of the camera are marked separately. The result is a series of 8 maps, corresponding to the 8 referenced photographs, in which every grid cell has one of three possible values characterizing cells which are snow-covered, snow-free or not visible from the camera.

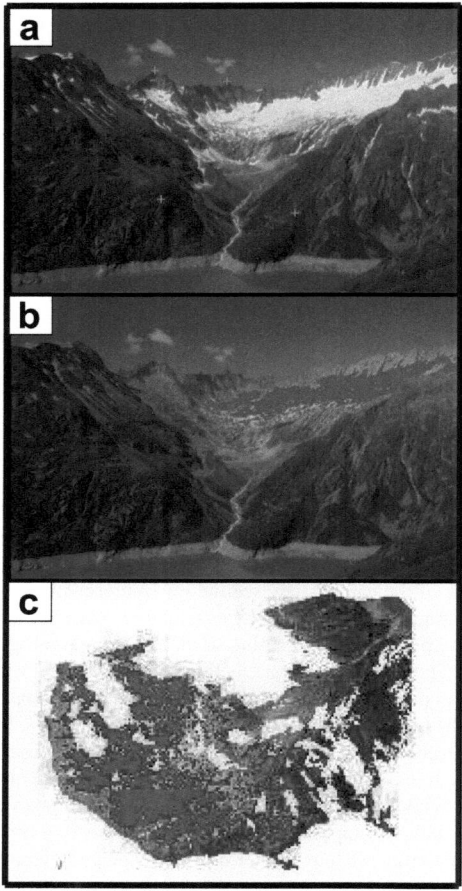

Figure 4.2: Steps of image processing. (a) Rotated and deskewed picture, GCPs are shown by crosses, (b) manually detected snow cover (red area), (c) georeferenced ortho-image of reflectance values (red: detected snow cover; green to gray: snow free; white: invisible for the camera). The example refers to the image taken on July 2, 2008.

4.3.2 Meteorological time series

The measurements of the AWS installed in the catchment are not continuous in time due to brief malfunctions of the temperature sensor and to the unheated rain gauge, which does not allow measurement of solid precipitation. The meteorological time series had, thus, to be adjusted with measurements from surrounding stations of the MeteoSwiss network.

For air temperature, the station at Grimsel Hospiz (13 km from study site) shows an almost perfect linear relation with the measurements of the AWS for both hourly (correlation coefficient $r^2 = 0.97$, average deviation $\overline{\Delta T} = 0.1\,°C$, standard deviation $\sigma_{\Delta T} = 1.4\,°C$) and daily ($r^2 = 0.99$, $\overline{\Delta T} = 0.1\,°C$, $\sigma_{\Delta T} = 0.8\,°C$) values. The temperature data of this station were, thus, shifted by the observed average deviation and adopted, completely replacing the time series collected at the AWS.

For choosing a representative precipitation station to adjust the precipitation time series, daily precipitation sums recorded at the AWS were correlated with those at surrounding MeteoSwiss stations. Since no solid precipitation could be measured at the AWS, only days with a minimum hourly temperature greater than 3 °C were considered. The correlations were computed using daily and three-day aggregated values. Precipitation is generally known as a parameter which correlates poorly, even over short distances, in alpine environments (e.g. Fliri, 1986; Frei and Schär, 1998). However, the correlation between the precipitation measured at the AWS and at the Göscheneralp station (3 km from the AWS, 280 m elevation difference), indicates that the approximation of the precipitation occurring in the catchment with that measured at the Göscheneralp station is reasonable. Daily aggregated values yield a correlation coefficient of $r^2 = 0.73$ ($\overline{\Delta P} = 0.3$ mm, $\sigma_{\Delta P} = 5.4$ mm) and three-day aggregated values of $r^2 = 0.87$ ($\overline{\Delta P} = 1.3$ mm, $\sigma_{\Delta P} = 5.5$ mm). In this case, daily precipitation sums at the AWS (P'_{AWS}) were generated by linearly adjusting the precipitation records from Göschneralp (P_{GOS}):

$$P'_{AWS} = \begin{cases} 1.31 \cdot P_{GOS} + 0.81 & \text{if } P_{GOS} > 0 \\ 0 & \text{if } P_{GOS} = 0 \end{cases} \quad (4.1)$$

The relations between daily values of air temperature and precipitation measured at the AWS (T_{AWS}, P_{AWS}) and the linearly transformed values of the chosen reference stations (T'_{AWS}, P'_{AWS}) are shown in Figure 4.3.

4.3.3 Melt modelling

The modelling of the snow depletion was performed with a distributed temperature-index model which accounts for potential direct clear-sky solar radiation and neglects refreezing (Hock, 1999). This temperature-index approach was chosen as it has been shown to be robust (e.g. Hock, 2005), is easy to implement and is driven by readily available meteorological data. The rate of daily snowmelt M_i at a given location i is computed according to:

$$M_i = \begin{cases} \left(f_M + r_{snow} \cdot I_{pot,i}\right) \cdot \overline{T}_i & \text{if } \overline{T}_i > 0°C \\ 0 & \text{if } \overline{T}_i \leq 0°C \end{cases} \quad (4.2)$$

where f_M is a melt factor, r_{snow} the radiation factor for snow, $I_{pot,i}$ the potential direct clear-sky solar solar radiation at the location i and $\overline{T}_i$ the mean daily air temperature at the same location. The potential direct clear-sky solar radiation is a function of the considered location and day of the year, accounting for effects of slope, aspect and shading, as well as for the seasonality in the incident solar radiation. Accounting for shading effects implies that the potential direct

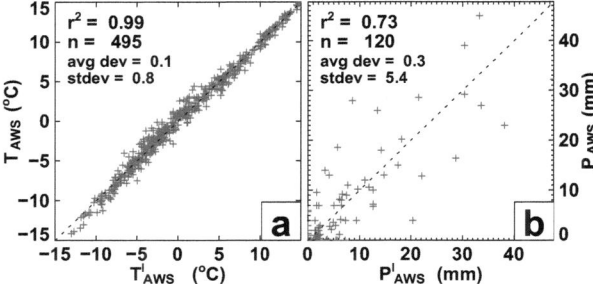

Figure 4.3: Relation between meteorological variables measured at the AWS (T_{AWS}, P_{AWS}) and linearly transformed data of the reference stations chosen (T'_{AWS}, P'_{AWS}). (a) Daily mean air temperature and (b) daily precipitation sum (days with minimum hourly temperature higher than 3 °C considered only). r^2: squared correlation coefficient, n: number of data points, *avgdev*: average deviation, *stdev*: standard deviation.

radiation is calculated at time steps shorter than one day. We used time steps of one hour and computed daily values by integrating the results over 24 hours. For further details refer to Hock (1999).

The mean daily air temperature at a given location is computed from the daily temperature at the AWS by means of a constant lapse rate. The lapse rate is determined from temperature records of 14 stations in the MeteoSwiss network ranging from elevations of 273 to 3580 m asl and is set to $-5.6 \cdot 10^{-3}$ °C m^{-1}.

The parameters in Equation 4.2 are adjusted, in order to match the long-term glacier volume changes inferred from the available DEMs (Bauder et al., 2007), with an iterative calibration scheme proposed by Huss et al. (2008a). We found $f_M = 1.30 \cdot 10^{-3}$ m (d °C)$^{-1}$ and $r_{snow} = 1.36 \cdot 10^{-5}$ m^3 (W d °C)$^{-1}$. Note that for the purpose of this study, only the parameters controlling snowmelt are of interest, although the used calibration procedure adjusts the parameters for icemelt as well.

4.3.4 Accumulation modelling

Precipitation is assumed to increase linearly with elevation with a lapse rate of dP/dz (Peck and Brown, 1962). The gauge catch deficit is accounted for with a correction factor c_{prec} (Bruce and Clark, 1981). A threshold temperature T_{thr} distinguishes snow from rainfall, whereas the fraction of solid precipitation r_s linearly decreases from 1 to 0 in the temperature range $T_{thr} - 1$ °C to $T_{thr} + 1$ °C (Hock, 1999).

The spatial variation in accumulation of solid precipitation is influenced substantially by the preferential deposition and redistribution of snow (e.g. Lehning et al., 2008). Since most redistribution processes are controlled by topographic effects, snow distribution patterns are often consistent over time (e.g. Elder et al., 1991; Luce et al., 1998). A simple way to take advantage of this fact was presented by Jackson (1994) which accounted for the complex patterns of snow redistribution through a predefined spatial precipitation matrix. The matrix is applied to every event of solid precipitation in the form of a location-dependent "snow multiplier" $D_{snow,i}$. The approach was applied successfully in some studies (e.g. Tarboton et al., 1995; Huss et al.,

4.3. METHODS

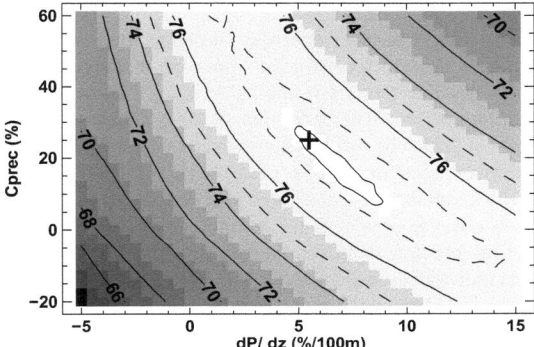

Figure 4.4: Calibration of the local precipitation gradient dP/dz and the correction factor accounting for the gauge catch deficit c_prec. The gray scale and the contour lines indicate the goodness-of-fit measure, defined as the percentage of grid cells visible from the camera for which a correct snow cover was calculated. The best parameter combination (goodness-of-fit measure = 78.3 %) is marked by a cross.

2009a).

Combining the described assumptions, the daily snow accumulation C_i at any location i with elevation z_i during a day with measured precipitation P_AWS at the AWS is calculated as:

$$C_i = P'_\text{AWS} \cdot (1 + c_\text{prec}) \cdot [1 + (z_i - z_\text{AWS}) \cdot dP/dz] \cdot D_\text{snow,i} \cdot r_\text{s}. \tag{4.3}$$

4.3.5 Calibration of the accumulation parameters

The calibration of the parameters of the accumulation model is performed in a multilevel iterative procedure. First, dP/dz and c_prec are adjusted. This is done by initializing the accumulation and melt model simultaneously at the end of August 2007 (a time of the year in which the catchment is assumed to be almost snow-free), and building up the snow cover during the 2007-2008 winter season. The simulated evolution of the melt-out pattern is then compared to the one observed during the melt season 2008. This procedure is repeated, systematically varying the parameters dP/dz and c_prec. The fraction of area seen by the fixed camera for which a correct snow cover is simulated was defined as goodness-of-fit measure. This measure, which can be interpreted as the degree to which the observed melt-out pattern can be reproduced correctly, is computed by comparing the calculated and observed snow cover in every grid cell and averaging the percentage of correctly calculated grid cells over the 8 snow cover maps available. As best estimate for the two parameters dP/dz and c_prec, the combination which leads to the highest goodness-of-fit measure is chosen (Fig. 4.4). We found dP/d$z = 5.5\%/100\,\text{m}$ and $c_\text{prec} = 25\%$.

In a second step, an analogous procedure is repeated systematically varying T_thr. The concomitant variation of dP/dz and c_prec with the systematic variation of T_thr showed that the best estimate for T_thr is almost independent from the choice of the first two parameters. The best matching of calculated and observed snow cover evolution was found for $T_\text{thr} = 1.2\,^\circ\text{C}$.

Finally, the snow multiplier matrix D_snow is adjusted in a third step, in order to optimize the spatial agreement between simulated and observed melt-out pattern. This is done iteratively, starting again the accumulation and melt modelling at the end of August 2007, and optimizing the previously defined goodness-of-fit measure for the melt-out patterns observed during the melt season 2008 (Fig. 4.5). The unity matrix is chosen as the initial estimate for D_{snow}. At the end of each iteration loop, which consists of a one-year model run starting in August 2007, a location dependent correction factor k is applied to D_{snow}. D_{snow} is assumed to be constant over the entire modelling period. The correction factor is calculated taking advantage of the information contained in grid cells for which a snow cover is simulated, but no snow cover is observed in the corresponding picture. In this case, the local value of k is chosen to be inversely proportional to the SWE simulated for the given grid cell. For grid cells where snow cover is observed, but no snow cover is simulated, the definition of a correction factor is more complicated since no direct information about the magnitude of the misfit is available. However, the time interval between individual snow cover maps allows to estimate how long the situation "no snow cover calculated, but observed" persists and thus, gives a measure for the magnitude of the misfit. In this case k is chosen to be inversely proportional to the length of the persistence of the situation during the melting season. Note that, since the 8 pictures define 7 time intervals in which the situation "no snow cover calculated, but observed" can occur (or not occur) independently, k assumes one of $\sum_{i=1}^{7} 2^i = 254$ possible discrete values. For grid cells where the snow cover is simulated correctly, or the fixed camera has no insight, the local value of k is set to 1. The adjustment of D_{snow} is performed in each iteration loop after the simulation of the last day with a snow cover map by multiplying D_{snow} with k. In order to cause D_{snow} to redistribute the solid precipitation without affecting the total amount, the matrix is normalized over the catchment domain to an average value of 1. After the adjustment of D_{snow} the melt and accumulation models are restarted, and the whole procedure repeated until convergence. The change in the goodness-of-fit measure compared to the value of the previous iteration loop is defined as convergence criterion. The iteration is aborted for a change less than 0.1 %.

Note that at this stage, it is impossible to discern whether the calibrated accumulation parameters are correct or biased in order to compensate errors arising form the melt calculation. This problem will be discussed and solved later in the text (see sections "Validation" and "Sources of uncertainty").

4.4 Results and discussion

4.4.1 Resulting snow accumulation distribution

The procedure for the adjustment of the snow multiplier matrix converges after about 10 iterations (Fig. 4.6a). On average over the 8 available snow cover maps, the snow cover can be reproduced correctly for about 93 % of the grid cells seen by the fixed camera (Fig. 4.6b). The high level of correctly calculated grid cells is almost constant over the melt-out season and independent from the degree to which the catchment is snow-covered (Fig. 4.6b).

The resulting snow accumulation distribution for a hypothetical day with 10 mm precipitation measured at the AWS and solid precipitation occurring in the entire catchment is shown in Figure 4.7. The effect of steep slopes is evident. On the upper ridges of the basin, as well as on the north-facing steep slopes and the lateral moraines of the forefield, snow accumulation is very limited. On the other hand, large snow deposits are recognizable in the southern sector of the dead-ice body. These deposits were observed indeed during the field survey and were mainly caused by avalanches. Note that in zones not seen by the camera and in zones which are snow covered throughout the melting season, the accumulation distribution remains a function

4.4. RESULTS AND DISCUSSION

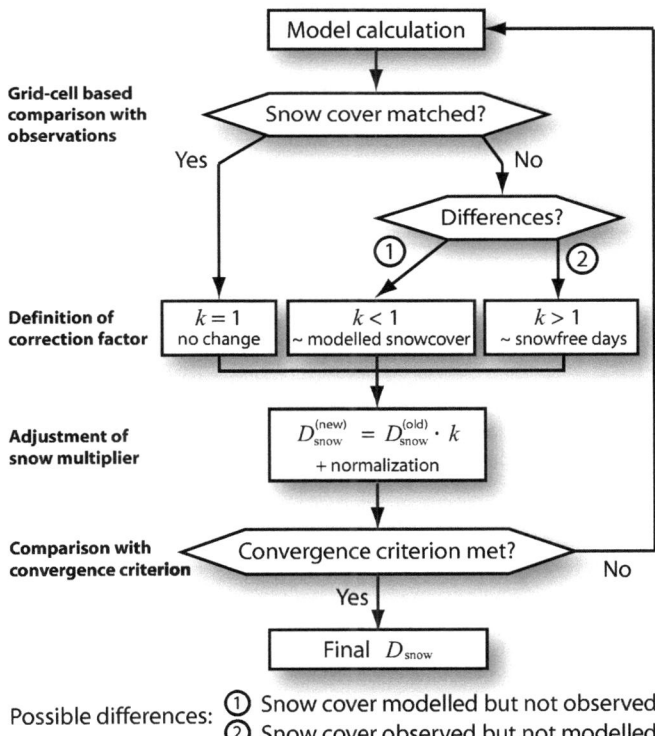

Figure 4.5: Calibration procedure used for adjusting the snow redistribution matrix D_{snow}.

48   CHAPTER 4. INFERRING THE SNOW ACCUMULATION DISTRIBUTION

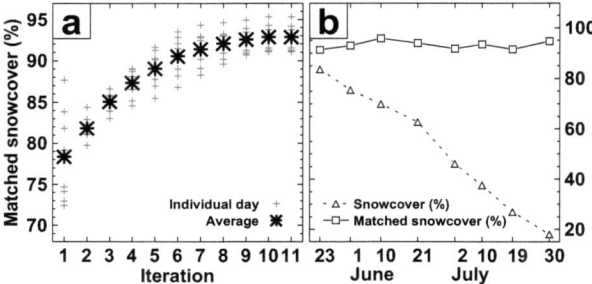

Figure 4.6: Evolution of the goodness-of-fit measure (percentage of snow-covered grid cells reproduced correctly) for the optimization procedure. (a) Convergence is reached after about 10 iterations. (b) The goodness-of-fit measure is almost constant over the melt-out period and is independent from the snow coverage observed in the catchment. Percentages are normalized over the portion of the catchment seen by the camera. Days of the year 2008 for which a snow-cover map is available, are shown on the time axis.

of elevation only, since no information is available for adjusting D_{snow} (Fig. 4.7, hatched area).

Since processes affecting the snow accumulation distribution are often controlled by topography (e.g. Elder et al., 1991), the values inferred for the snow multiplier matrix D_{snow} are correlated with two topographic variables: local slope (Fig. 4.8a) and curvature (Fig. 4.8b). The correlations were evaluated only for zones for which D_{snow} could be adjusted (non-hatched area in Fig. 4.7).

Curvature is a topographic attribute used as proxy for the degree of wind exposure of a given grid cell and has often been used for explaining some variance in the observed distribution of snow accumulation due to wind drift (e.g. Elder et al., 1989; Blöschl et al., 1991; Huss et al., 2008a; Carturan et al., 2009). We compute the parameter as the difference between the local altitude and the average altitude of a portion of terrain lying inside a given radius (Carturan et al., 2009). According to this definition, negative curvature values indicate concave terrain (hollows) and positive values convex terrain (hills).

In spite of the large scatter, values inferred for D_{snow} show a significant correlation with local slope ($r^2 = 0.54$, Fig. 4.8a). This is attributable to gravity driven mechanisms of snow redistribution as avalanching and slushflow, as well as wind-drift and -erosion processes taking place on steep walls because of downward winds. The linear relation fitted to the data indicates that the terrain is almost permanently snow-free for local slopes exceeding about 55 degrees. This is in agreement with results of earlier studies (e.g. Witmer, 1984; Blöschl and Kirnbauer, 1992). Based on the intercomparison of different studies, Blöschl et al. (1991) pointed out that the local slope for which hillsides become permanently snow-free may depend on the cell size of the underlying grid. In order to analyze this effect, we reevaluated the correlation between inferred values of D_{snow} and local slope resampling the DEM to 50, 75 and 100 m resolution. For increasing cell size we observed a decrease in the correlation coefficient ($r^2 = 0.54$ for 25 m and $r^2 = 0.44$ for 100 m cell size) as well as a decrease in the local slope for which hillsides become permanently snow-free (55° for 25 m and 45° for 100 m). This is attributed to the smoother topography caused by larger grid cells, resulting in reduced local slope.

The linear correlation between D_{snow} and curvature is weak ($r^2 = 0.09$, Fig. 4.8b). This can

4.4. RESULTS AND DISCUSSION

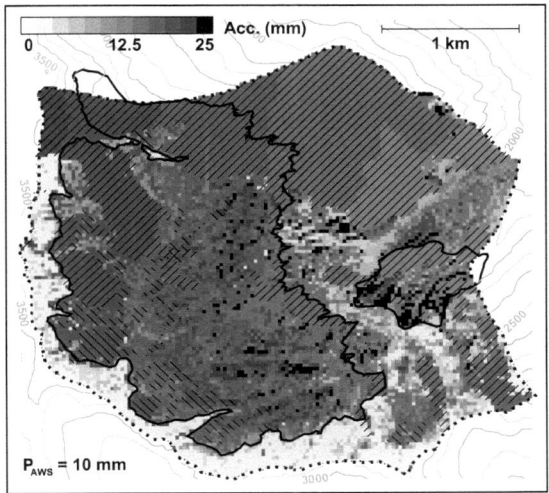

Figure 4.7: Snow accumulation distribution inferred for a hypothetical day with 10 mm precipitation at the AWS and no liquid precipitation occurring in the catchment. In zones not seen by the fixed camera (hatched ///) and in zones with permanent snowcover (hatched \\\), D_{snow} cannot be adjusted.

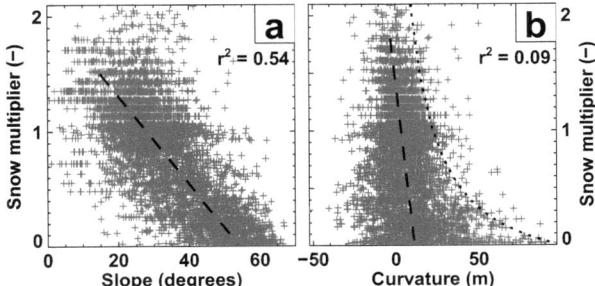

Figure 4.8: Correlation of the snow multiplier D_{snow} with (a) local slope, and (b) local terrain curvature. Dashed lines are fitted by linear regression. The dotted line in (b) indicates a threshold curvature for the maximal predicted value of D_{snow}. Slope is evaluated on a 25 m grid, curvature with a radius of 125 m.

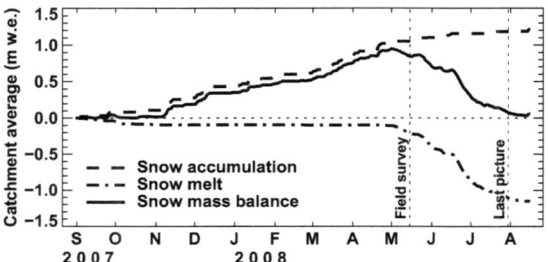

Figure 4.9: Temporal evolution of the mass balance of the snowpack. A time span of one year is shown. The day of the field survey and the day for which the last snow-cover map is available are marked. The mass balance components are averaged over the catchment area seen by the fixed camera.

however not be interpreted as the absence of influence of curvature on D_{snow}. The effect may be masked by interactions with other factors or not be linear. In fact, curvature controls the maximal value inferred for D_{snow}. Very low values of D_{snow} are found over almost the whole range of curvature values, whereas very high values of D_{snow} occur only at curvatures close to zero (dashed line in Fig. 4.8b). This means that zones where snow accumulation is significantly below average occur independently of the concavity of the terrain, whereas zones with high above-average snow accumulation are only present in rather flat areas (neither concave nor convex terrain). The sensitivity of the definition of curvature was assessed varying the evaluation radius in the range of 50 to 200 m. The effect on the relation between curvature and D_{snow} is small. The correlation coefficient becomes somewhat larger towards a smaller radius ($r^2 = 0.20$ for 50 m radius and $r^2 = 0.08$ for 200 m radius), but the limiting effect of curvature remains visible.

The linear correlation of D_{snow} with some other topographically-controlled parameters was explored without detecting significance or simple interpretable patterns. In particular, this was the case for aspect and potential solar radiation. However, similarly as stated for curvature, the absence of significant correlations can not be interpreted as the absence of influence of the analyzed factors on D_{snow} as effects may be masked by interactions with other parameters.

4.4.2 Validation

In order to validate the inferred snow accumulation distribution, the model was re-initialized at the end of August 2007, building up the snow cover during the 2007-2008 winter season. Thereby, all solid precipitation events are redistributed according to D_{snow}. According to the modelling, the field survey providing the direct SWE measurements was conducted two weeks after the maximal SWE was reached in the catchment (Fig. 4.9). Considering a one-year period starting from the end of August 2007, 87 % of the accumulation and 16 % of the melt occurred prior to the field survey. The collected SWE measurements are thus suitable for a validation of the inferred snow accumulation distribution and the accumulation model, as they are not biased significantly by melt processes. The calculated and measured SWE distribution for May 15, 2008 agrees well in the point-to-point comparison (Fig. 4.10a). The difference between observed and calculated average snow depth $\overline{\Delta h}$ is 5 cm w.e., with a standard deviation of the residuals $\sigma_{\Delta h}$ of 28 cm w.e.. The inferred snow accumulation distribution explains 51 %

4.4. RESULTS AND DISCUSSION

of the observed variance in the sample of measured snow depths.

In order to compare the accuracy of the resulting snow distribution with other possible methods for estimating the SWE distribution, two additional experiments were conducted.

(1) In a first experiment, the SWE distribution for the day of the field survey was calculated building up the snow cover during the 2007-2008 winter season using the same accumulation and melt model as described above, but setting the snow multiplier matrix D_{snow} uniformly to 1. The residuals between calculated and measured snow depths at the locations of direct measurement were then used as estimator for the accuracy of the experiment. The standard deviation of the residuals was $\sigma_{\Delta h}$=31 cm w.e. and is, thus, not significantly larger than $\sigma_{\Delta h}$ calculated when D_{snow} is adjusted as previously described. This relatively small standard deviation can be explained with the correlation of the measured snow depth with altitude ($r^2 = 0.42$) and the fact that setting the snow multiplier matrix to 1 causes the precipitation to become a function of altitude only (Eq. 4.3). The difference between mean measured and mean calculated SWE is, however, high ($\overline{\Delta h} = -33$ cm w.e.). This offset could be corrected by adjusting the factor c_{prec}, but this would cause the degree to which the observed melt-out pattern can be reproduced to decrease from about 93 % (Fig. 4.6) to about 77 % (Fig. 4.4).

(2) In a second experiment, the SWE distribution for the day of the field survey was obtained by inverse distance weighting of the direct measurements. In this case, the uncertainty was assessed by starting the inverse distance interpolation N times, omitting one measurement point each time. The residuals were then computed, storing the deviation between measured and calculated SWE at the omitted measurement point in each step ("cross validation"). The standard deviation of the residuals of this experiment is $\sigma_{\Delta h}$=26 cm w.e., whereas the average snow depth is matched exactly ($\overline{\Delta h} = 0$ cm w.e.). The explained variance is 50 % (Fig. 4.10b). In respect to the variance of the residuals, no significant difference is observed between the results of this second experiment and the SWE distribution inferred through the adjustment of D_{snow}. The result is astonishing, since the inverse distance weighting procedure is based exclusively on the snow depths measurements, whereas the inferred accumulation distribution for solid precipitation relies on the observed melt-out pattern only. Of course, the variance of the residuals of the inverse distance weighting interpolation is expected to decrease with increasing number of snow depth measurements and with a more homogeneous distribution of the measurement points. This, however, is a result which is difficult to achieve, since the required effort in terms of field work is large.

A comparison of the two described experiments shows that (a) the adjustment of D_{snow} with the presented procedure enhances significantly the degree to which the measured SWE distribution can be reproduced, and (b) the inferred SWE distribution is comparable in terms of standard deviation of the residuals to the result achieved by interpolating direct snow depth measurements with a inverse distance interpolation scheme.

Once the accumulation model has been proven to be sufficiently accurate, the performance of the combined model (accumulation and melt) during the ablation season can be interpreted as well. According to Fig. 4.9, only 12% of the accumulation but as much as 78% of the melt of a one-year period starting from the end of August 2008 occurs during the day of the field survey (May 15, 2008) and the last available snowcover map (July 30, 2008).The matching of the modelled melt-out pattern evolution with the observed one in this period thus indicates that the melt model performs well and that the melt parameters calibrated using the long-term ice volume changes are in a reasonable range.

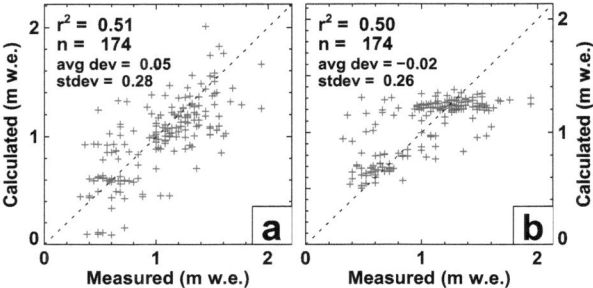

Figure 4.10: Comparison between snow depths measured on May 15, 2008 and calculated snow depth using (a) the inferred accumulation distribution, and (b) an inverse distance interpolation procedure applied N times, omitting 1 data point at each time. For the comparison, only measurements in regions seen by the fixed camera were considered (Fig. 4.1).

4.4.3 Sources of uncertainty

Different sources of uncertainty affect the accuracy of the inferred parameters for the accumulation model; estimating their magnitude is, however, difficult.

The accuracy of the image-referencing procedure was assessed by picking one prominent feature on all available images, for which the position was determined from the aerial photograph used for the preparation of the DEM. The feature was chosen as far as possible from the position of the fixed camera, where the resolution of the image in terms of image pixels per real meter length is minimal (about 0.41 pixel per meter), and in a location where the incidence angle of the camera was estimated to be representative for the whole picture. The images on which the feature was picked were then georeferenced and the transformed position of the feature compared with the previously determined coordinates. In all cases the transformed position matched the known coordinates within twice the resolution of the DEM, i.e. 50 m. We assume this value to be representative for the average accuracy of the georeferencing procedure, although for flat zones, where the incidence angle of the camera becomes small, the accuracy may be lower.

The accuracy of the compiled snow cover maps is also determined by the accuracy with which snow-covered areas of the picture can be detected. Quantifying this accuracy is difficult, as it is both location- and time-dependent. Snow detection is relatively easy in the early season, when almost the whole catchment is snow-covered and the area becoming snow-free is concentrated in the low-lying glacier forefield (where the contrast to the bare ground is high) and on steep mountain ridges (where the incidence angle of the camera is high). On the other hand, detection becomes difficult later in the season, when the snowline retreats to the glacierized parts of the basin, where the contrast between snow and underground (i.e. bare ice) is less pronounced. Based on the experiences collected during the manual detection of the snow cover on the images, we estimate the accuracy to be in the order of 50 to 75 m in the glacierized parts of the catchment and in the order of the DEM resolution (i.e 25 m.) in the remaining area.

Several uncertainty sources are introduced due to assumptions made for accumulation and melt modelling. The daily precipitation sums driving the accumulation model are not measured in situ but taken from a station outside the catchment. The threshold temperature distinguishing solid and liquid precipitation is assumed to be constant, although Rohrer (1989) reported

4.4. RESULTS AND DISCUSSION

a tendency towards lower threshold temperatures in summer than in winter, and constant lapse rates are assumed for precipitation and temperature too. The threshold temperature is based on daily mean temperature, excluding thus smaller snowfall events, e.g., during cold nights, and the snow multiplier matrix, redistributing the solid precipitation instantaneously, is assumed to be constant over time. Finally, the parameters of the melt model are calibrated with long-term ice volume changes and assumed to be constant over time as well, although a significant year-to-year variation was pointed out by Huss et al. (2009b). Assessing the effect of each assumption on the overall accuracy is difficult, as no direct measurements are available for validation. Moreover, uncertainties induced by different assumptions can potentially compensate for one another. The inferred snow multiplier matrix could, thus, also be regarded as a parameter lumping corrections for uncertainties arising from different model assumptions, although being mainly determined by processes controlling the redistribution of snow accumulation. However, the timing of the field survey and the beginning of the snowcover monitoring divide two distinct periods dominated either by accumulation or by melt. This allows us to state, that both model components - representing accumulation and melt - reproduce the observations reasonably and independently from each other.

The direct measurements of snow depth used for validation are affected by uncertainties as well. Although the accuracy in individual snow depth measurements is high (accuracy of reading = ± 5 cm), uncertainty is introduced by the roughness of the underground, which may cause individual measurements to not be representative for larger areas. On glaciers, the ice surface is generally smoother than the surface of bare soils, thus making individual measurements representative for larger portions of terrain, but the detection of the correct sounding horizon is not always easy, especially for measurements in the firn area. Moreover, a homogeneous distribution of the collected measurement (Fig. 4.1) was hampered by the avalanche risk in the catchment. This resulted in a data set which is slightly biased toward relatively small surface slopes.

Further uncertainty is introduced by the conversion of measured snow depth into SWE. Also in this case, the snow density measured at a single location may not be representative for the whole area. This uncertainty source has potential for systematic errors when assessing the total SWE of the catchment. The modelling procedure has the advantage of computing SWE directly, without requiring further information on snow density.

4.4.4 Potential, recommendations and limitations for further applications

The method as presented relies on images taken through terrestrial oblique photography in the visible band (VIS) and was developed in an Alpine environment. However, it can easily be applied using any kind of sensor able to observe melt-out pattern evolution (e.g. aerial photography or satellite imagery) and is not restricted to glacierized or mountainous catchments. We see a large potential for application to remote and inaccessible areas, where the seasonal depletion of the snowcover has significant impacts on hydrology or other processes (e.g. Siberia, Himalaya, Canadian plains).

For a faster and more accurate snow detection in the images, we recommend the use of imagery sensible to both, VIS and near infrared (NIR), as combination of this information has been shown to be suitable for automatic snow detection (e.g. Hall et al., 1995). The accuracy of the snow detection also depends on the incidence angle of the fixed camera. The position of the camera should, thus, be chosen so that the incidence angle is as close as possible to 90°. We furthermore recommend the use of a digital camera, which reduces the effort required for operating the camera and simplifies some steps of the image processing. When working on VIS, attention has to be payed to the time of the day at which images are taken, since shadows or reflections may compromise the possibility of detecting snow visually.

We calibrate the parameters of the melt model to long-term ice volume changes. This is, however, not the only possible procedure. Empirical temperature-index snow-melt parameters for different areas may be, for example, calibrated with snow-depth data collected at stakes or by a sonic ranger and in combination with local temperature and precipitation measurements.

The monitoring of the melt out pattern should start the latest at the beginning of the melting season. This is crucial for allowing the melt and the accumulation model to be validated separately.

According to our experience, snow cover maps compiled at intervals of 10–20 days are suitable for the application of the method. However, the interval at which individual pictures are taken should be considerably smaller (e.g. one to three days) as, especially when operating in VIS, several pictures may not be suited for further processing because of reduced visibility due to fog or clouds. Local meteorological conditions may thus limit the applicability of the method to some regions.

The calibration scheme for D_{snow} (Fig. 4.5) limits the application to areas where the seasonal snowcover melts completely, or at least to regions where snow layers from different seasons are clearly recognizable. This, for instance, is often not the case in accumulation zones of glaciers.

4.5 Conclusions

The evolution of the melt-out pattern observed during one ablation season using conventional oblique photography was combined with a temperature-index melt model and a simple accumulation model to infer the snow accumulation distribution of a small Alpine catchment. On average over the melt-out season, the observed snow cover depletion could be reproduced correctly in 93 % of the area seen by the fixed camera. The inferred snow accumulation distribution was validated with in situ snow depth measurements. The comparison with the results of an inverse-distance interpolation of direct measurements showed that the inferred accumulation distribution is able to explain almost the same fraction of variance in the snow depth measurements. Correlations with topographic variables showed a large scatter, but a significant linear relation with local slope was found. Curvature was detected as a factor controlling the maximal local accumulation. Quantifying an overall accuracy is difficult as there are various sources of uncertainty which may compensate for one another to some degree.

Our results suggest that the method is suitable for inferring the spatial distribution of SWE in remote areas without the need of direct access for field measurements, as long as melt-out patterns are observable. As the distribution of snow in mountain regions is highly complex and difficult to reproduce with unconstrained modelling, the presented method has the potential to enhance the performance of models in different fields. Better representation of SWE and its spatial variability is of crucial importance for distributed hydrological models providing runoff forecasts, the monitoring of glacier mass balance or avalanche forecasting.

Acknowledgments

Financial support for this study was provided by the BigLink project of the Competence Center Environment and

4.5. CONCLUSIONS

Sustainability (CCES) of the ETH Domain. Swisstopo provided topographic maps and aerial photographs. H. Bösch evaluated the DEMs from aerial photographs. Many thanks to J. Corripio, who provided substantial support in the georeferencing of the images. M. Lehning made valuable comments on earlier versions of the manuscript. S. Braun-Clarke edited the English. We acknowledge M. Anderson for the editorial work and R. Dadic and two anonymous reviews for their helpful remarks.

Chapter 5
Conclusions and outlook

In this thesis, two new methods for inferring two variables of difficult access – ice-thickness and snow-accumulation distribution – were presented and the potential for a wide application shown. However, the approximations used lead to limitations in the accuracy and further improvements are possible.

For the method for estimating the ice thickness distribution (Chapter 2), the sensitivity of the method with respect to the input parameters, and in particular, to the correction factor C, was shown. A careful choice of the parameters is, thus, essential.

As explained in Section 2.5, C is introduced to account for (a) the uncertainties in the flow-rate factor A; (b) the approximation of the shear stress distribution by a linear relation; (c) the influence of the basal sliding; (d) the approximation of the specific ice volume flux at the center of the profile with the mean ice volume flux across the profile.

Uncertainties in the flow-rate factor A arise since it has been shown that A departs from the values determined in laboratory experiments when used in flow-modeling experiments (e.g. Hubbard et al., 1998; Gudmundsson, 1999). Deviations up to a factor of two have been found. The approximation of the shear stress distribution by a linear relation was proposed by Nye (1965) and a linear relation has also been proposed to link the basal sliding to the deformation velocity (e.g. Gudmundsson, 1999). The approximation of the specific ice volume flux at the center of the profile with the mean ice volume flux across the profile is an approximation required in order to avoid assumptions of the bedrock shape when computing the specific ice-volume flux.

While the influence of the approximation (d) can be shown to have a minor effect (Section 2.4), addressing the approximations (a) to (c) in more detail has the potential of constraining C in a more physically based manner and thus improving the accuracy of the method. In fact, using C as a tuning parameter which lumps different kinds of uncertainties voids the parameter from most physical meaning.

When leaving the method unchanged, two types of data seem promising for better constraining the value of the parameter:
First, the potential of the ice-thickness measurements, collected during numerous field campaigns throughout the Swiss Alps, is not exhausted yet. A more sophisticated analysis may help to derive rules to infer C from morphological characteristics. Currently, it is suggested that C may depend upon the glacier type, having different values for valley and cirque glaciers for instance, but the relation is not explored in a quantitative way yet.
Second, surface velocities could be used to constrain C too. A three-step procedure, which seems promising, could adopt the following scheme: (a) compute an ice-thickness distribution

with a first estimate of C; (b) compute surface velocities using the estimated ice-thickness and compare the results to measured velocities; (c) adjust the ice-thickness distribution by adapting the local value of C in a way that the measured surface velocities are matched.

When adopting this scheme, two options may be of interest for computing surface velocities. One approach would be to consider an approximation for individual flowlines, as the shallow-ice approximation. In this case, the proposed three-step procedure may even be simplified as C would be fixed in a deterministic way. A second approach could be to calculate surface velocities with a more sophisticated approach, e.g. with a 3D model solving the full Stokes equations as proposed by Jouvet et al. (2008, 2009). In this case a stepwise procedure seems necessary, as the problem is non-linear (Jouvet et al., 2008). The result would be a self-consistent bedrock topography regarding ice-thickness and surface velocities, which, up to now, is not necessarily the case.

From the theoretical point of view, a second, iterative procedure, seems possible to infer C. Again, an initial guess of C would be necessary to compute a first estimate of the ice-thickness. From that, the basal shear stress along the flowlines – which corresponds to the denominator in Equation 2.7 (i.e. $(C\rho g \sin\overline{\alpha})^n$) multiplied by the local ice thickness – could be computed and used to calculate a new, local ice thickness. This would then fix the local value of C.

The implementation of this promising procedure was attempted but, unfortunately, not successful. The reason may be the following: when computing the local basal shear stress, the local surface slope is required. In Equation 2.7 however, only the average surface elevation of the glacier appears, as the local slope is accounted for in a subsequent step (see point "F" on page 9). This causes a conflict which results in the non-convergence of the algorithm. A solution should address this point first. Note that the reason for implementing the average (and not the local) slope in Equation 2.7 is, that in doing so, the sensitivity of the exact position of the individual flowlines in respect to the calculated ice thickness can be significantly reduced. In the other case, the position of the flowlines would be determinant for the computed local slope and, thus, for the computed ice thickness on the flowline. Since this ice thickness is then interpolated over the whole glacier (see point "E" at page 9), the estimate of the total ice volume would be (strongly) influenced as well.

When a 3D flow model as proposed by Jouvet et al. (2008, 2009) is available, a third iterative procedure could be used for improving the estimated ice-thickness distribution. The main idea would be to iteratively adjust the ice-thickness distribution in order to achieve consistency between the estimated bedrock topography, the calculated flow field and the current glacier geometry. Starting with an initial guess for the ice-thickness distribution, the flow model could be initialized and run one time-step forward. The reaction of the modeled surface topography should then give hints regarding the plausibility of the ice-thickness distribution: in regions of the glacier where the adjustment of the surface appears unrealistic, the local ice-thickness should be corrected. The procedure seems promising since unrealistic rates of surface-elevation change can be detected and direct measurements of surface-elevation change could be used to further constraining the problem.

Finally, the way the local surface slope α is accounted for (point "F" at page 9) may be improved too. Currently, α is accounted for through a factor proportional to $(\sin\alpha)^{\frac{n}{n+2}}$. As one notes immediately, the factor tends to ∞ as α tends to 0 and so the local ice thickness does. To avoid the problem, a filtering of the surface slope was introduced. However, no rule has been provided for avoiding an arbitrarily choice of the value with which the slope is filtered. Following Kamb and Echelmeyer (1986), α should be averaged in flow direction over an interval of about 4 times the local ice thickness. This procedure is easily applicable when individual flowlines are considered, but is difficult to implement in a distributed way. In fact, this would require the computation of the iceflow direction at any point of the glacier, as well as an iterative procedure, since the local ice thickness is initially unknown. Currently, the choice of the value of the filter has an effect on how many "overdeepenings" are produced in the calculated bedrock topography. One should be aware of the problem when interpreting such features, especially when trying to forecast the formation of new glacier ponds and lakes, once the glacier has retreated.

When the estimate of the total ice volume of the Swiss Alps (Chapter 3) was presented, the attention of the wide public – including media – was concentrated on the quantification of the relative volume loss in the last decade, rather than on the estimated total volume. In this respect it has to be noted, that the estimation of the volume loss was performed in a very simple way, that is, applying a mean time series of mass balance and letting the area of the individual glacier unaltered (see Section 3.4). Since the mean time series was derived averaging individual series of 30 glaciers and the total number of analyzed glaciers is large (about 1500, see Table 3.3), this may be defensible from a statistical point of view. However, the fact that the mass balance of individual glaciers can vary significantly even over short geographical distances (Huss et al., 2010a) and that glaciers are currently experiencing a significant retreat, indicates that the approach should be revised at least in part. The first step for an improved estimate would be to apply the averaged time series only to glaciers where no glacier-specific time series is available. Since for most of the largest glaciers such a series has been compiled, the uncertainty of the total volume loss should decrease significantly, as the total volume – and thus, the relative change as well – is mainly determined from these glaciers. The second step would be to account for the evolution of surface area, occurred during the considered period. For larger glaciers, a suitable approach to address the issue may be the parametrization scheme of glacier retreat as proposed by Huss et al. (2010b). For smaller glacier, the assumption of an unchanged surface area should have a very minor influence on the overall picture.

For a more precise estimate of the total volume, emphasis has to be put in the largest glaciers. Up to now, the data basis for estimating the bedrock topography in the basin of Grosser Aletschgletscher is scarce. The acquisition and evaluation of field data from this glacier should have priority since, as shown, the basin contains almost one quarter of the total ice volume in the Swiss Alps.

Large potential was shown for the method presented for inferring the snow accumulation distribution from time-lapse photography (Chapter 4) as well. Similarly, however, some detail may be readdressed in more detail.

The method was applied to one catchment (Damma) and for one season only (summer 2008). It would be necessary to assess the performance of the method in different circumstances. Apart from the performance of the method itself, interesting questions would arise from the comparison of the results from different catchments and between different years. The most important one would be, to assess how stable the observed melt-out pattern and snow-accumulation distribution is through time. For the Damma catchment, which would be a suitable starting point, the data basis for 2009 is now available.

The control of topographic variables on the observed patterns was addressed only in an approximative way. The performed analysis is limited to two simple linear regressions. An analysis with multiple linear regression or some geostatistical approach should guarantee further insights in the controlling mechanisms. In this respect, the set of more than 1400 snow-depth measurements collected during the Damma field campaign on April, 2009 (see Appendix B) gives a rather unique possibility for explorative analysis.

The method was applied by manually detecting the snowcover from each individual picture. As mentioned in Section 4.4.4, this was necessary since the camera was operated in the visible band only, and no robust automatic procedure exist for detecting snow on that band. For an application on a regional scale however, an automated detection of the snowcover seems indispensable. Working with a camera sensible to the IR band would be a significant improvement.

Finally, the method seems suitable for estimating the parameters controlling the mass balance distribution in the method presented for estimating the ice thickness. The combination of the two methods seems an exciting idea.

The objective of this thesis was to provide simple methods for inferring two variables of difficult access, as the ice-thickness and accumulation distribution, from more readily available data. I hope and have the impression, that this goal was achieved. However, when using the methods, I encourage to stay critical and to try to further improve them, rather than having the impression, that the "world formula was found", all problems are solved and no field measurements are longer required.

Appendix A

Ice thickness measurements on glaciers in the Swiss Alps

A.1 Overview

First experiments in measuring the ice thickness of glaciers in the Swiss Alps go back to the late 1920s. Using seismic techniques, H. Mothes tried to measure the ice thickness at the Konkordiaplatz of the Grosser Aletschgletscher in 1929. The adopted methods were developed in 1927 already (Mothes, 1927).
Early studies, based on seismic measurements too, were performed on Rhonegletscher in 1931 (Gerecke, 1933; Jost, 1936) and on Unteraargletscher starting from 1936 (Mercanton, 1936). Seismic methods remained the commonly used measurement technique until the 1950s. Measurement campaigns take further place in the region of the Konkordiaplatz (Mothes, 1929; Haefeli and Kasser, 1948), but extended to other glaciers too. Measurements were performed in 1943 on Vadret da Morteratsch (Kreis, 1944; Renaud and Mercanton, 1948), in 1946 on Glacier de la Plaine Morte (Renaud and Mercanton, 1948; Kreis et al., 1947), starting from 1948 on Gornergletscher (Renaud and Mercanton, 1948; Süsstrunk, 1950) and in 1949 on Glacier du Mont Collon (Süsstrunk, 1951) and Zmuttgletscher (Süsstrunk, 1950).
From the 1950s on, drilling campaigns became more popular for measuring ice thickness and seismic methods were applied more sparsely, e.g. in 1958 at Konkordiaplatz (Thyssen and Ahmad, 1969) or 1959 on Findelgletscher (Süsstrunk, 1959). First experiments with thermal- and hydrothermal drilling were started on Gornergletscher already in 1948, in the frame of the studies for the realization of the Grande Dixence. Drilling campaigns were started in 1949 on Zmuttgletscher, 1958 on Findelgletscher (Swissborung, 1959), 1960 on Glacier du Trient (VAWE, 1961), 1966 on Glacier du Giétro and Glacier du Tournelon Blanc (Kasser and Aellen, 1974), 1972 and 1975 on Grubengletscher (Kasser and Aellen, 1976; Kasser et al., 1982; Röthlisberger, 1979), 1973 on Oberaletschgletscher (Kasser and Aellen, 1976) and starting from 1977 on Unteraargletscher and Glacier du Brenay (Kasser et al., 1983). In many cases the focus of the measurements were not the ice thickness directly. At Colle Gnifetti for example, core drilling campaigns started in 1976 already (Oeschger et al., 1977; Schotterer et al., 1981), but the glacier bedrock was reached first only 6 years later.
In the first half of the 1960s, some experiment was started to determine the ice thickness using geoelectric techniques, e.g. in 1963 on Unteraargletscher and 1964 on Ewigschneefeld (Röthlisberger and Vögtli, 1976). This technique was used 1982 on the Grubengletscher as well, but was never employed extensively.

First experiments using radio-echo soundings were performed in 1978 (Thyssen, 1979). The encouraging experiences led to more extensive campaigns between 1980 and 1982 on Colle Gnifetti, Glacier de la Plaine Morte, Findelgletscher, Grubengletscher and Rhonegletscher (Haeberli et al., 1982), during which the technique was further developed. The technique became state of the art from the 1980s on. Extensive measurement campaigns were performed in 1982 on Glacier de Saleina, Allalingletscher (VAW, 1983a) and Griesgletscher (VAW, 1983b), between 1987 and 1998 on Unteraargletscher (Bauder et al., 2003), 1988 and 1998 on Glacier du Corbassière (VAW, 1998), 1990 on Haut Glacier d'Arolla (Sharp et al., 1993), 1990, 1991 and 1995 on Grossen Aletschgletscher, 1993 on Claridenfirn (Funk et al., 1997), 1997 on Glacier du Giétro (VAW, 1998), 2001 and 2002 on Vadret da Morteratsch (research work of AWI), 2001 and 2003 on Triftgletscher (Müller, 2004), 2002 on Ghiacciaio del Basodino (Bauder et al., 2006), 2003 and 2008 on Rhonegletscher (Zahno, 2004; Farinotti et al., 2009b), 2004, 2005 and 2007 on Gornergletscher (Huss, 2005; Riesen, 2006; Eisen et al., 2009), 2006 on Unteren Grindelwaldgletscher (VAW, 2007b), 2006 and 2007 on Glacier de Zinal (VAW, 2006, 2007a; Farinotti, 2007; Huss et al., 2008b) and 2007 on Silvrettagletscher (Farinotti et al., 2009b).

In 2008, an experiment with an airborne radio-echo sounding system was performed on Gornergletscher. The technique showed the potential for performing ice thickness measurements in an extension and a speed which was unachievable until now: through the use of an helicopter, measurements which would have required days and days of fieldwork can be collected in a few hours. The system was further applied in 2008 on Findelgletscher, Allalingletscher, Schwarzberggletscher, Rhonegletscher and Triftgletscher, as well as in 2009 on Grossen Aletschgletscher but the results for Trift-, Rhone- and Grosser Aletschgletscher are not yet completely evaluated (status May 2010).

A.2 Detailed information

An overview of the ice thickness measurements collected in the Swiss Alps until 2009, as well as of the concerning publications and reports, is given in the following table.

Table A.1: Overview of ice thickness measurements collected in the Swiss Alps until 2009. *Glacier*: Name of the glacier; *Year*: Year in which the measurements were collected; *Meth*: Method of measurement (1a = radio-echo sounding, 1b = airborne radio-echo sounding , 2a = hydrothermal drilling, 2b = mechanic drilling, 3 = seismic, 4 = geoelectric, 5 = hydrometric); *dd*: Availability of digital data (x = yes, o = no, p = partially); *Literature*: Publications and reports concerning the measurements; *Notes*: further information about the measurements. In the notes, "pr" stands for "profile(s)", "bh" for "borehole(s)".

Glacier	Year	Meth	dd	Literature	Notes
Aletsch	1947	3	o	Haefeli and Kasser (1948)	1 pr below Konkordiaplatz
	1996	1a	x	Farinotti et al. (2009a)	7 pr between Konkordiaplatz and tongue, deepest points not reached
	2009	1b			Results non yet available (May 2011)
- Konkordiaplatz	1929	3	p	Mothes (1927, 1929)	

Continued on next page

A.2. DETAILED INFORMATION

Table A.1 – Continued from previous page

Glacier	Year	Meth	dd	Literature	Notes
	1958	3	p	Thyssen and Ahmad (1969)	
	1990-91	2a	x	Hock et al. (1999)	2 bh to the bed, 1 ca. 20-30 m above
	1990-91	1a	x	Farinotti et al. (2009a)	7 pr, deepest points not reached
- Ewigschneefeld	1964	4	o	Röthlisberger and Vögtli (1976)	
	1976-77		o	Atlas der Schweiz, Bl. 80, (Kasser et al., 1982, 1983; Oeschger et al., 1977)	Core drillings
Allalin	1982	1a	x	VAW (1983a)	3 pr at ca. 2800 m asl
	2008	1b	x	VAW (2009)	Pr until ca. 3500 m asl
Arolla, Haut	1990	1a	x	Sharp et al. (1993)	Good coverage, poor accuracy
Basodino	2002	1a	x	Bauder et al. (2006)	5 pr
Brenay	1977	5	o	Kasser et al. (1983)	
Clariden	1993	1a	x	Funk et al. (1997)	4 cross-pr, 1 longitudinal-pr at ca 2900 m asl
Corbassière	1988	1a	x	VAW (1998)	5 pr in tongue region
	1998	1a	x	VAW (1998)	5 pr above, 1 below the steep step
Fiescher	2000	1a, 2b	p	Zweifel (2000)	8 pr, 1 bh near Fieschersattel
Findel	1958-59	2a+b	o	Swissborung (1959)	
	1959	3	o	Süsstrunk (1959)	
	1980, 82	2a	p		Research work VAW (A. Iken)
	1980, 82	1a	o	Haeberli et al. (1982)	Research work VAW
	2008	1b	x		2 pr, poor coverage
Giétro	1966	3, 2a	o	Kasser and Aellen (1974)	
	1997	1a	x	VAW (1998)	4 cross-pr, 1 longitudinal-pr
Gorner	1948	3	o	Renaud and Mercanton (1948)	
	1948-49	3	o	Süsstrunk (1950)	
	1948-50	2a	o	Archive Grande Dixence	
	1969	2b	o	Bezinge et al. (1969)	Commited by Grande Dixence
	1974, 79	2a	o	Kasser et al. (1982)	
	2004	1a	x	Huss (2005)	10 pr below Gornersee
	2005	1a	x	Riesen (2006)	8 pr below Gornersee
	2004-08	2a	x	Sugiyama et al. (2008); Eisen et al. (2009)	Several bh until bedrock
	2008	1b	x		Pr below 3500 m asl
- Colle Gnifetti	1976-77	2b	o	Oeschger et al. (1977)	Core drilling
	1978	1a	o	Thyssen (1979)	
	1980-81	1a	x	Haeberli et al. (1982)	
	1980, 82	2b	o	Schotterer et al. (1981)	Core drilling. Bedrock hit in 1982
Gries	1982	1a	o	VAW (1983b)	Bedrock mapping near tongue
	1987-89	1a	x	VAW (1983b); Vieli et al. (1997)	2 pr, research work VAW (M. Funk)
	1999	1a	x	von Deschwanden (2008)	14 pr

Continued on next page

APPENDIX A. ICE THICKNESS MEASUREMENTS

Table A.1 – Continued from previous page

Glacier	Year	Meth	dd	Literature	Notes
Gruben	1972	2b	o	Kasser and Aellen (1976)	
	1975	2a	o	Kasser et al. (1982); Röthlisberger (1979)	
	1981	1a	o	Haeberli et al. (1982)	Pr at tongue, until 3000 m asl
	1982	2a, 4	o		Research work VAW (W. Haeberli)
Mont Collon	1949	3	o	Süsstrunk (1951)	
Morteratsch	1943	3	o	Kreis (1944); Renaud and Mercanton (1948)	
	2001-02	1a	x		Research work AWI (P. Huybrechts, O. Eisen)
Oberaletsch	1973	2a	o	Kasser and Aellen (1976)	
Pizol	2010	1a	o	Huss (ress)	25 pr
Plaine Morte	1946	3	o	Renaud and Mercanton (1948); Kreis et al. (1947)	
	1980	1a	p	Haeberli et al. (1982)	First attempts with VAW-device, 1 measurement point
	2010	1a	o		17 pr Research work Uni Fribourg (M. Huss)
Rhone	1931	3	o	Gerecke (1933); Jost (1936)	
	1980	1a	o	Wächter (1979); Haeberli et al. (1982)	
	2005-09	2a	p		Several bh to the bed
	2003	1a	x	Zahno (2004)	12 pr
	2008	1a	x	Farinotti et al. (2009b)	8 pr
	2008	1b	p		Good coverage, especially at the tongue, poor accuracy
Rossboden	1997	1a,3,4	p	Oberholzer and Salami (1998)	Tongue only, good coverage
Saleina	1982	1a	o		Research work VAW (B. Ott)
Schwarzberg	2008	1b	x	VAW (2009)	Good coverage
Sex Rouge	2010	1a	o		2 pr Research work Uni Fribourg (M. Huss)
Silvretta	2007	1a	x	Farinotti et al. (2009b)	9 pr
Theodul	2002	1a	x	VAW (2002)	3 pr above 3000 m asl
Tournelon Blanc	1966	2a	o	Kasser and Aellen (1974)	
Trient	1960	2a	o	VAWE (1961)	1 pr at 2070 m asl
Trift	2001, 03	1a	o	Müller (2004); VAW (2004)	5 pr below, 1 above the steep step
	2008	1b	o		Results non yet available (May 2010)
Tsanfleuron	1999	2a	o	Hubbard et al. (2000)	5 bh near tongue

Continued on next page

Table A.1 – Continued from previous page

Glacier	Year	Meth	dd	Literature	Notes
	2010	1a	o		10 pr Research work Uni Fribourg (M. Huss)
Tschierva	2005	1a	o	Joerin et al. (2008)	Pr below 2500 m asl
Unteraar	1936-50	3	o	Mercanton (1936); Renaud and Mercanton (1948)	Measurements in 1936-39,47,48,50
	1963	4	o	Röthlisberger and Vögtli (1976)	
	1977, 81	2a	o		Research work VAW (A. Iken)
	1987-91	1a	x	Bauder et al. (2003)	21 pr, most below junction
	1997-98	1a	x	Bauder et al. (2003)	10 and 9 pr on Finsteraar- and Lauteraar-branches, respectively
	2000	1a	x	Bauder et al. (2003)	8 pr on Strahlegg-branch
U. Grindelwald	2006	1a	x	VAW (2007b)	6 pr at tongue, poor quality
Zinal	2006-07	1a	x	Farinotti (2007); Huss et al. (2008b)	12 pr on main branch
Zmutt	1949	3	o	Süsstrunk (1950)	
	1949	2a	o	Archive Grande Dixence	
Zupó	2002	1a	p		8 pr near F. dal Zupó; research work VAW (M. Lüthi)

A.3 Maps

The maps on the following pages show the location of the ice thickness measurements (circles) for glaciers for with digital data are available at VAW - ETH Zurich. The year to which the glacier outlines refers to is stated in the figures. Contour lines inside the glacier boundary correspond to the glacier bed generated with the methods described in Chapter 2. Contour lines outside the glacier show the surface topography according to the Swisstopo DHM25.

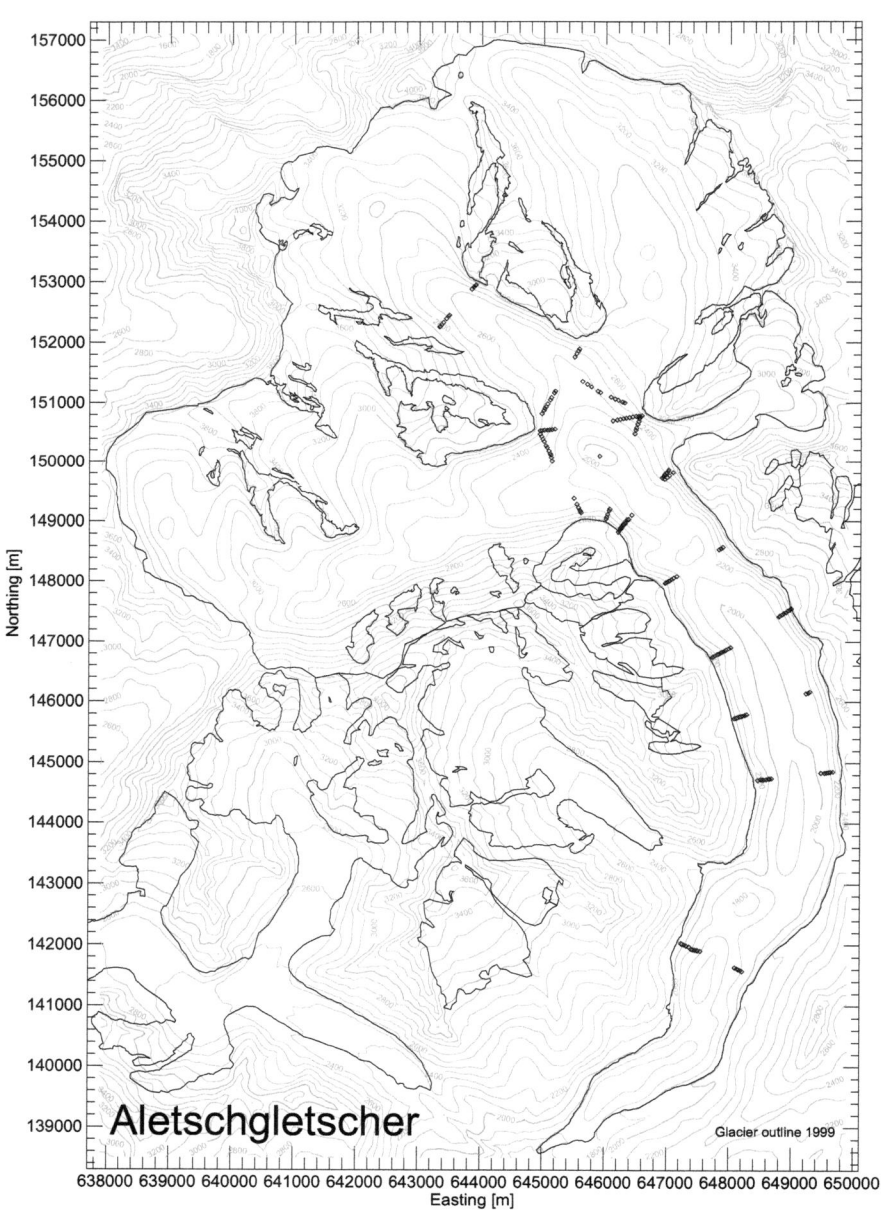

A.3. MAPS

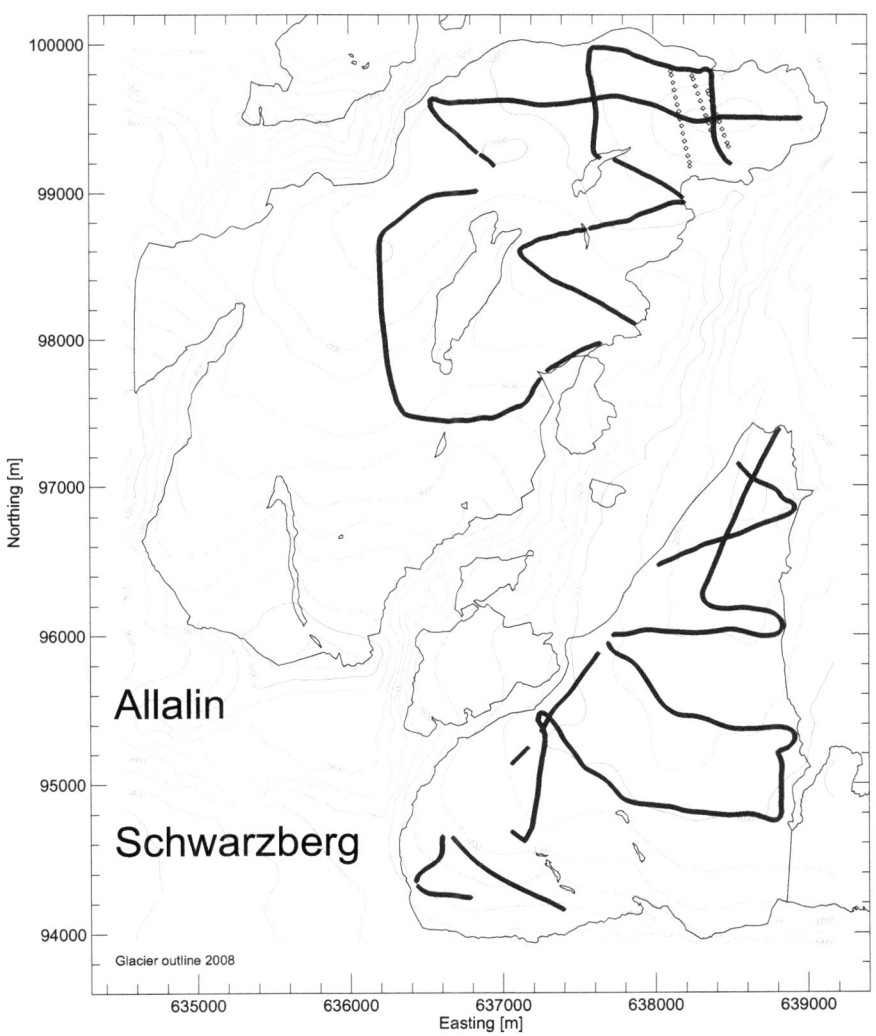

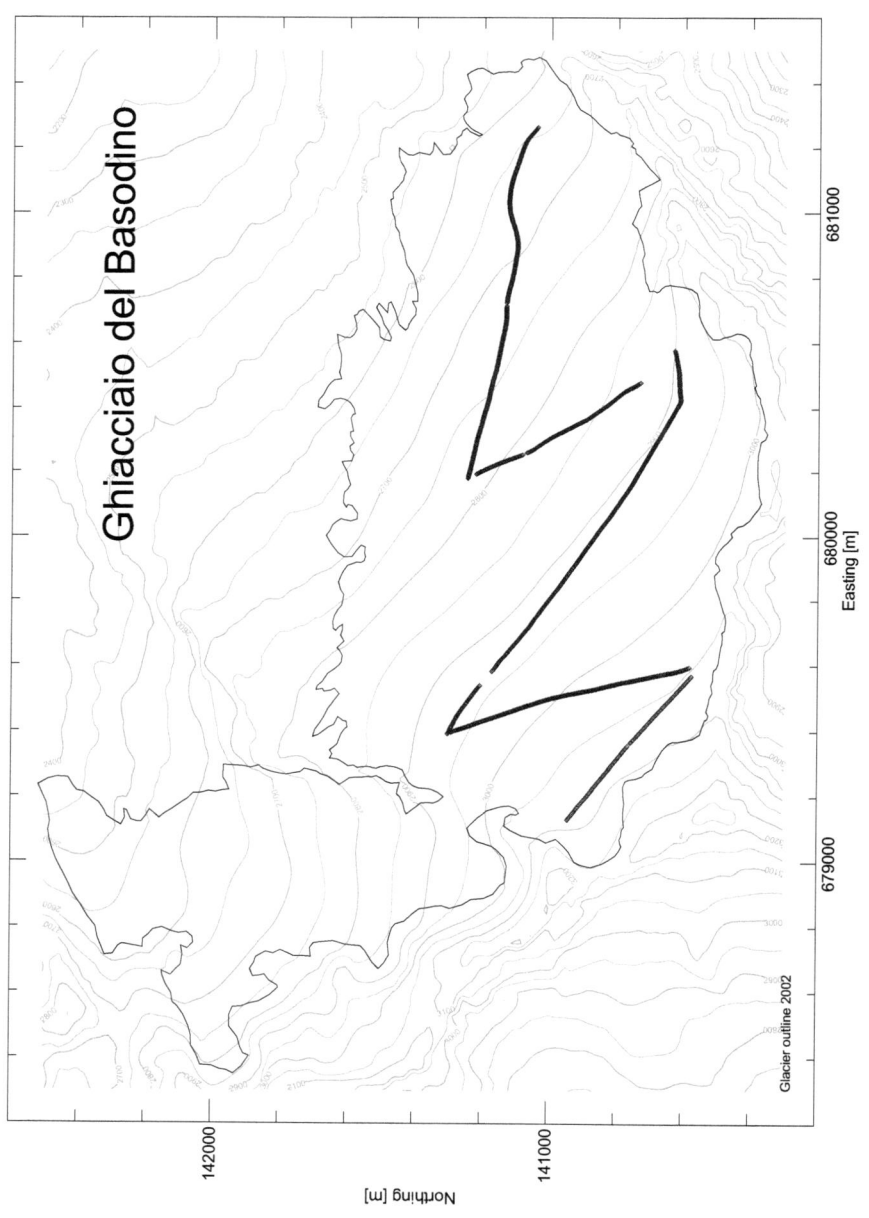

A.3. MAPS

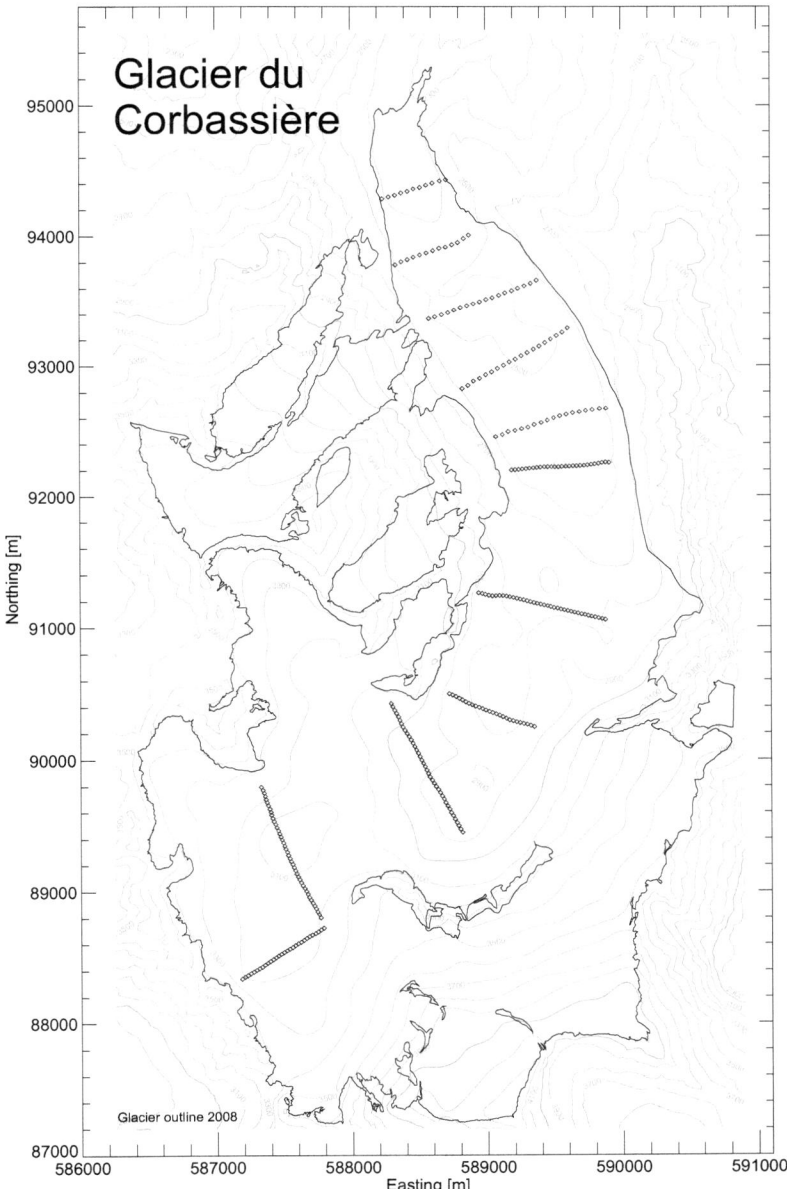

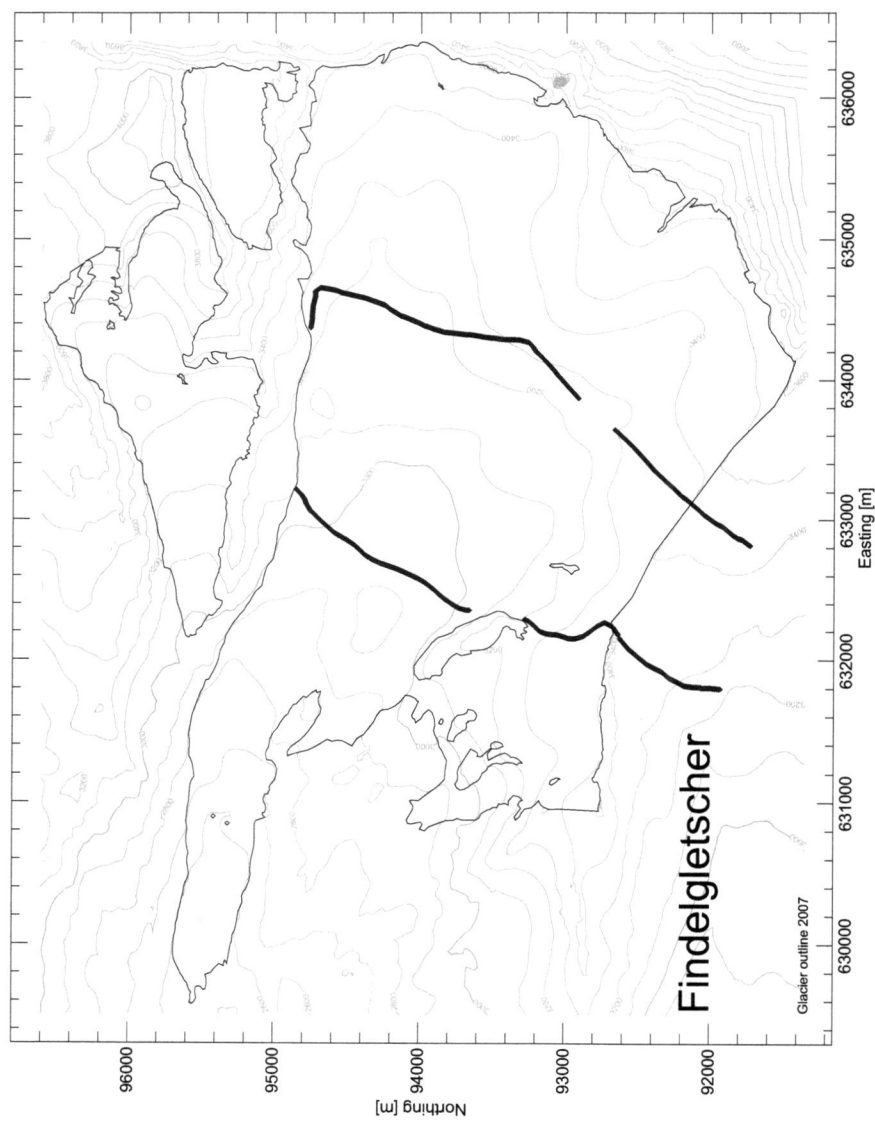

A.3. MAPS

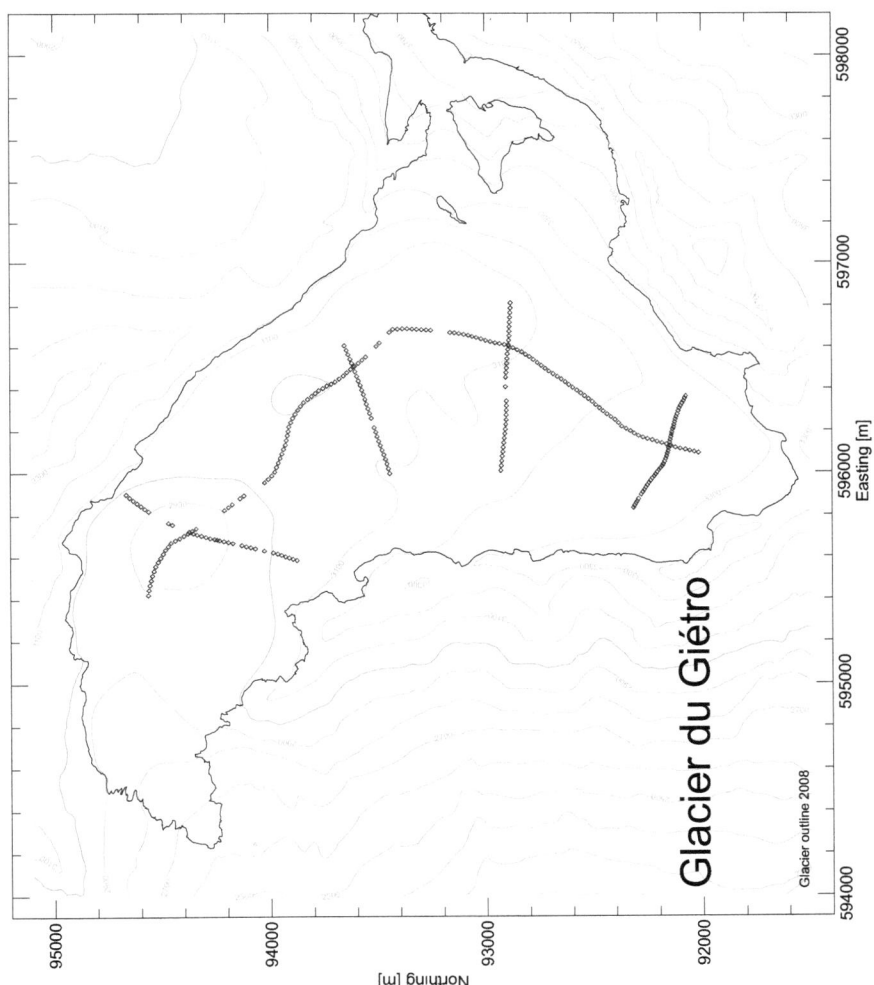

APPENDIX A. ICE THICKNESS MEASUREMENTS

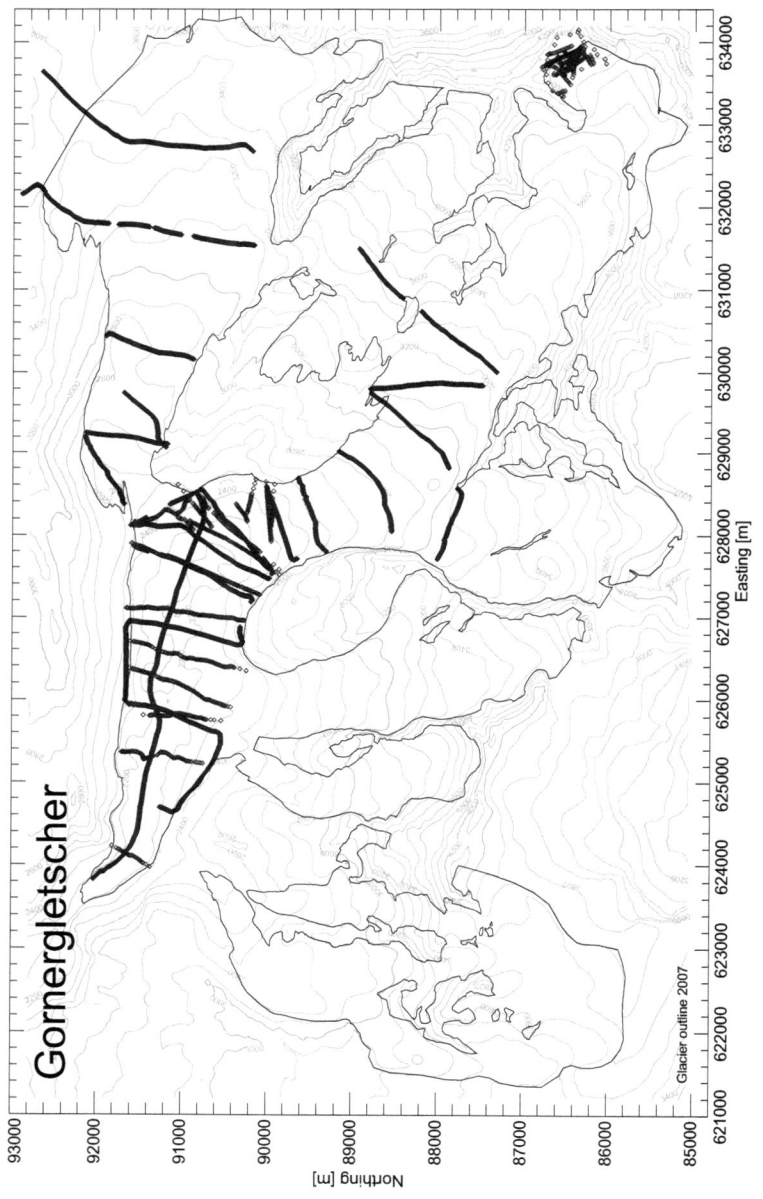

A.3. MAPS

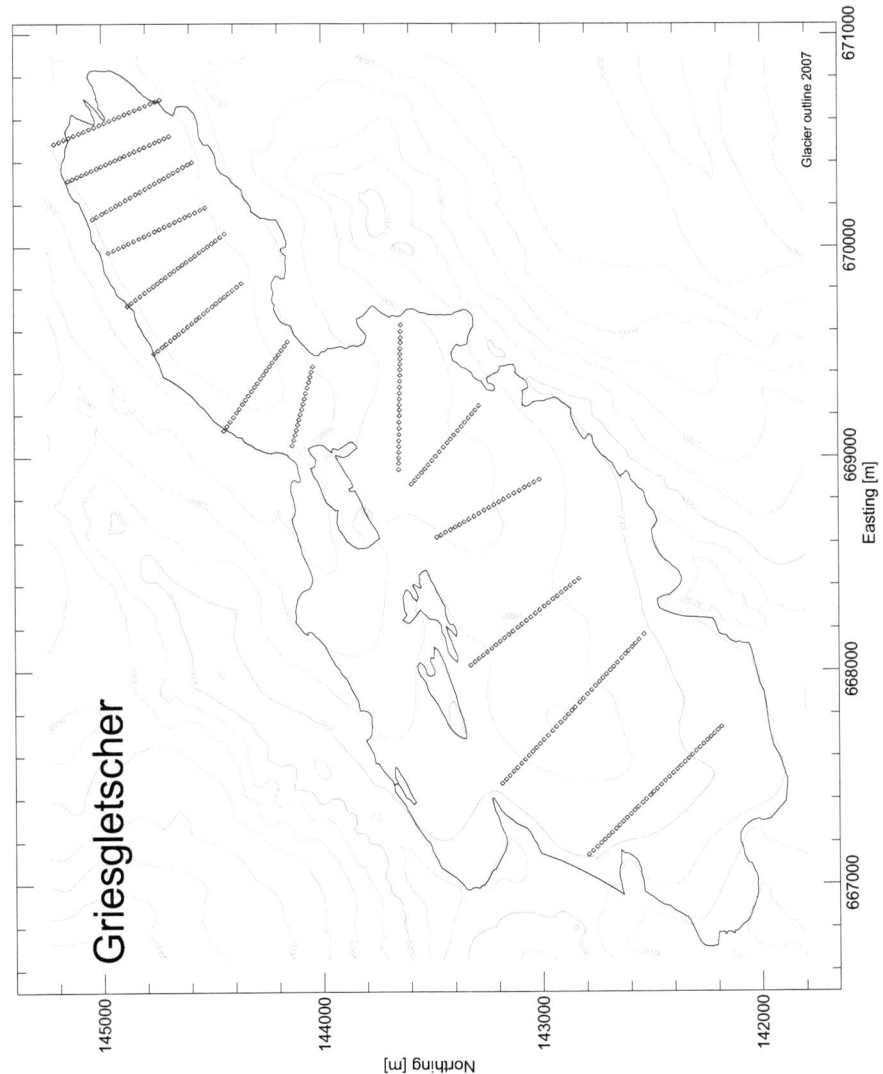

APPENDIX A. ICE THICKNESS MEASUREMENTS

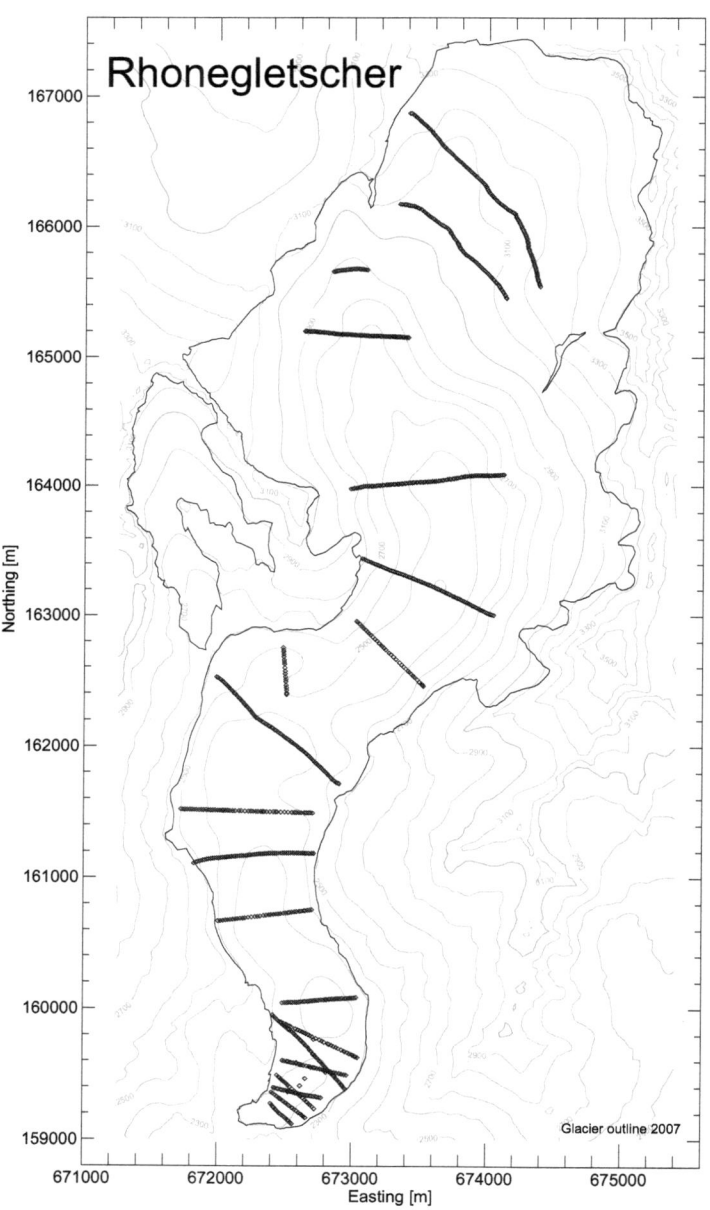

A.3. MAPS

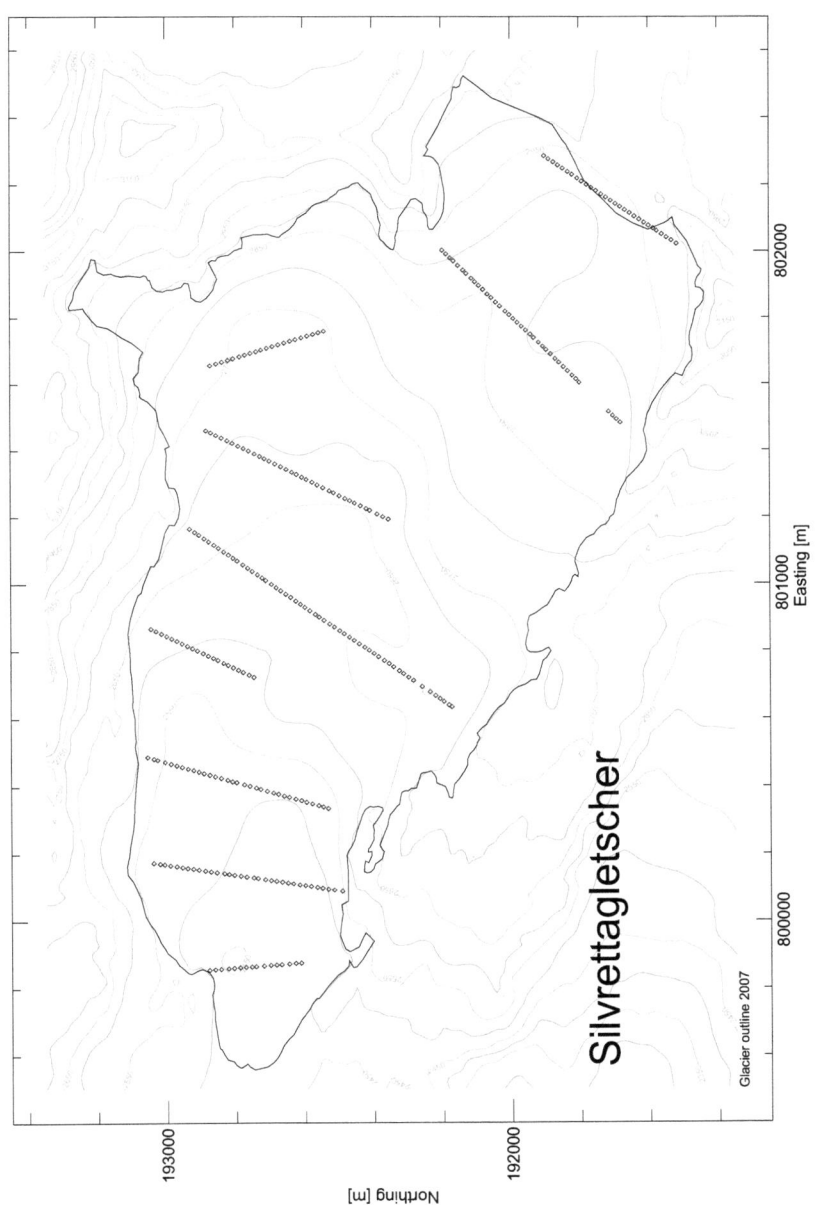

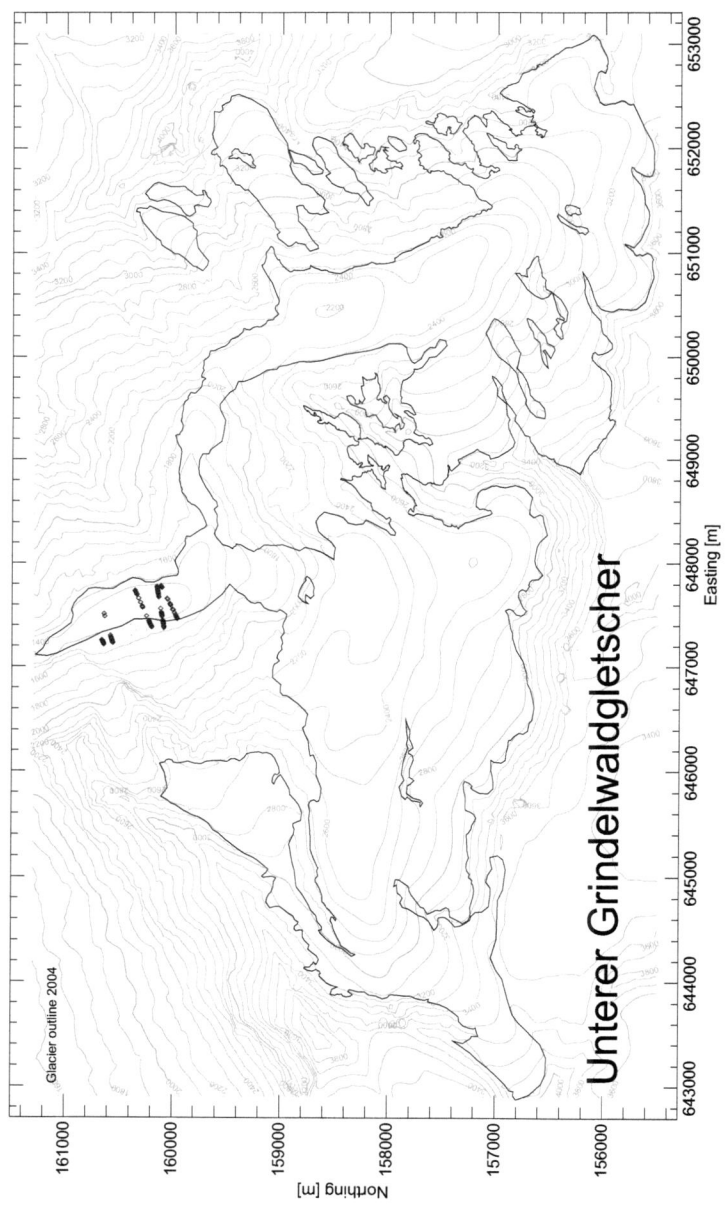

A.3. MAPS

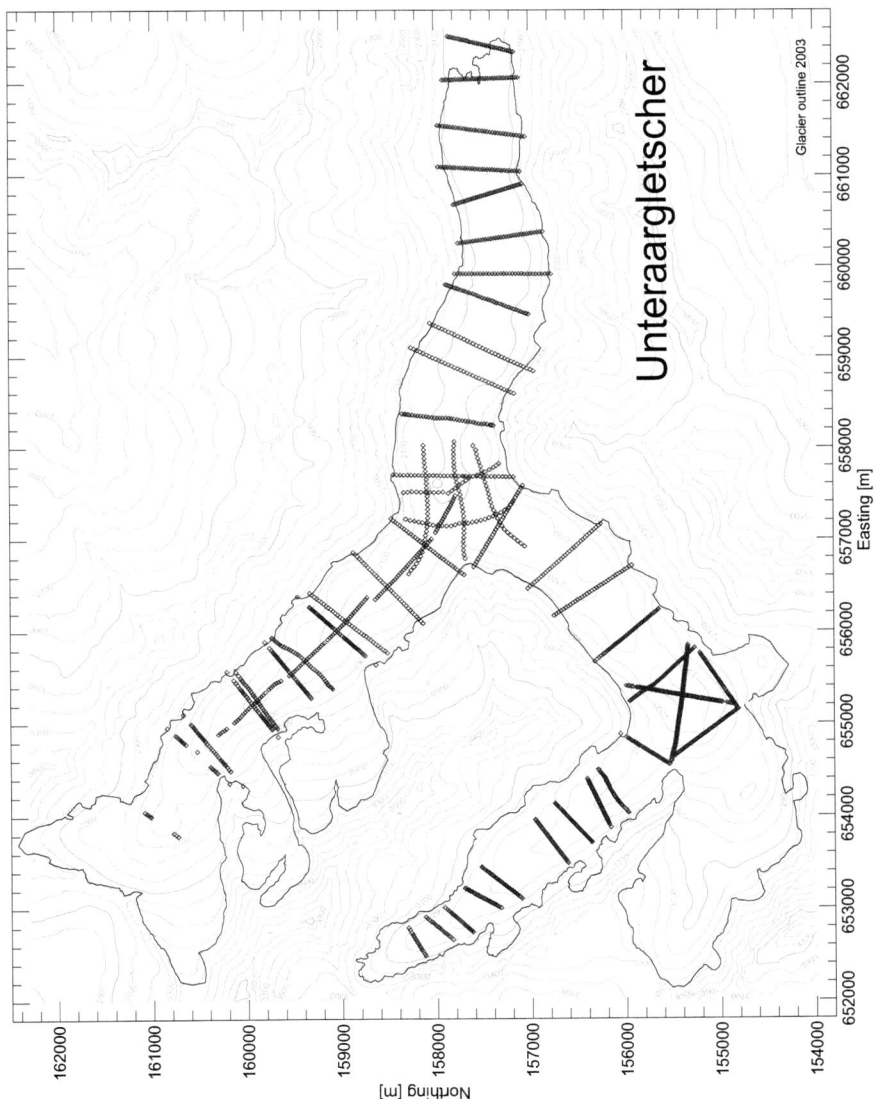

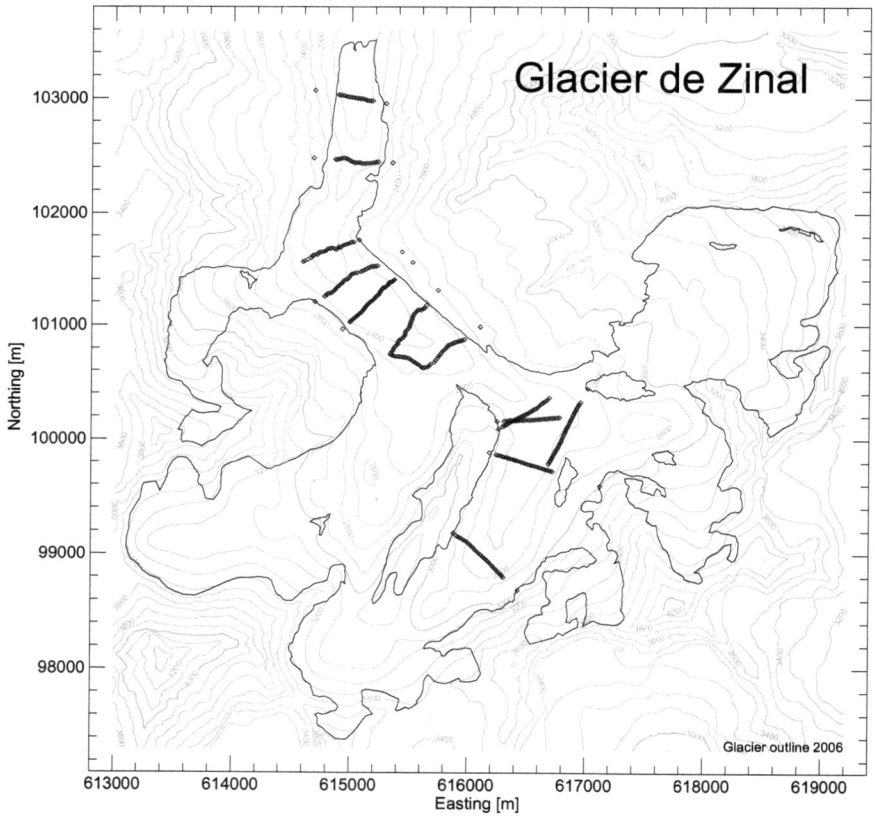

Appendix B

Snow-depth measurements on Dammagletscher, April 2009

During the period from April 6 to April 8, 2009 a large snow-depth measuring campaign was conducted on Dammagletscher. This was the second snow-depth measuring campaign after May 15, 2008 (see Chapter 4). During three days, eight people from the institutes VAW-ETH Zurich and WSL-SLF Davos collected more than 1400 snow-depth measurements and determined the snow density in three different snow-pits.

The data give a rather unique possibility for explorative analysis and are available at VAW. The map on Figure B.1 displays the locations of the individual measurements.

Figure B.1: Location of the measurements collected during the field campaign on Dammagletscher in April 2009. Snow-depth measurements are shown by crosses, snow pits for snow-density measurements with squares.

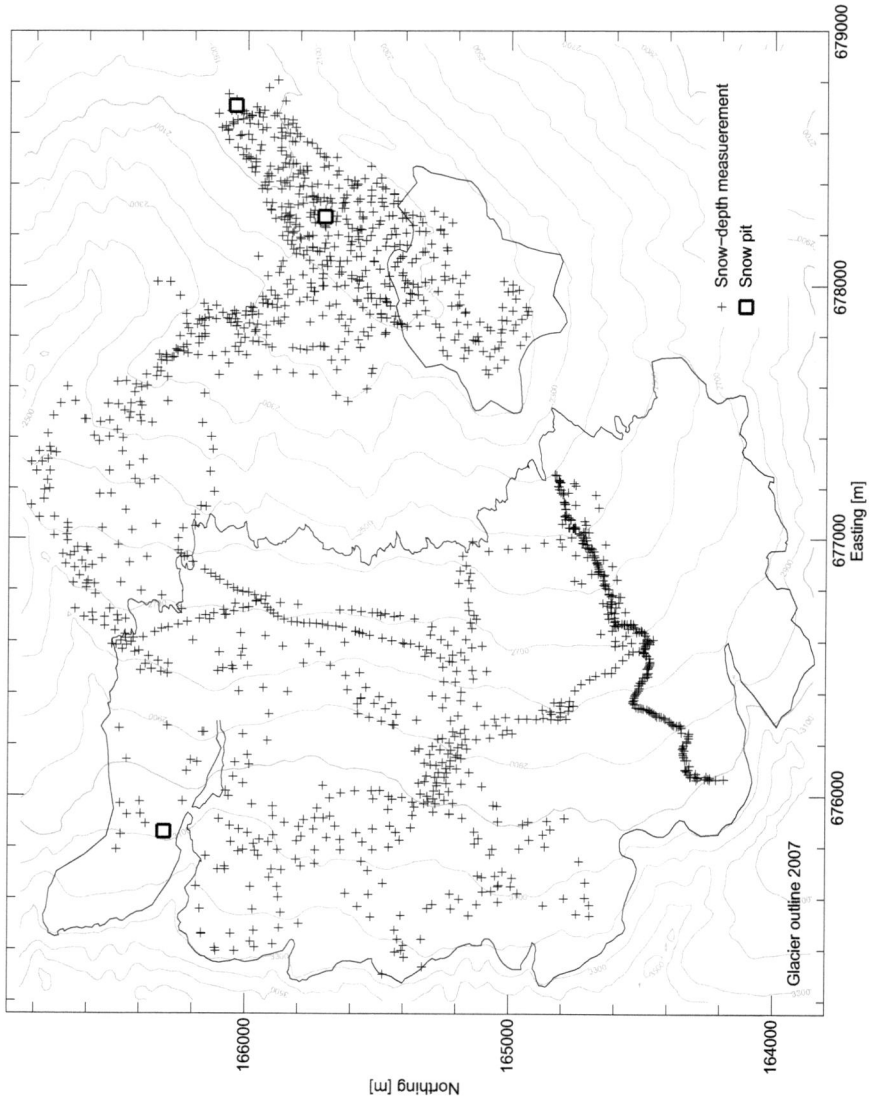

Bibliography

Agassiz, L. (1840). *Etudes sur les glaciers*, volume 1. Jent & Gassmann, Neuchâtel.

Arendt, A. A., Echelmeyer, K. A., Harrison, W. D., Lingle, C. S., and Valentine, V. B. (2002). Rapid wastage of alaska glaciers and their contribution to rising sea level. *Science*, 297(5580):382–386.

Bahr, D. B., Meier, M. F., and Peckham, S. D. (1997). The physical basis of glacier volume-area scaling. *Journal of Geophysical Research*, 102(B9):20355–20362.

Basist, A., Bell, G. D., and Meetemeyer, V. (1994). Statistical relationships between topography and precipitation patterns. *Journal of Climate*, 7:1305–1315.

Bauder, A., Eisen, O., and Kappenberger, G. (2006). Il ghiacciaio del Basodino. In *La misurazione dei ghiacciai in Ticino - Estratto dal quaderno "Dati" N.2 giugno 2006 dell'ufficio di statistica del Cantone Ticino*, pages 29–31. Dipartimento del Territorio, Divisione Ambiente, Sezione forestale.

Bauder, A., Funk, M., and Gudmundsson, G. H. (2003). The ice thickness distribution of Unteraargletscher (Switzerland). *Annals of Glaciology*, 37:331–336.

Bauder, A., Funk, M., and Huss, M. (2007). Ice volume changes of selected glaciers in the Swiss Alps since the end of the 19th century. *Annals of Glaciology*, 46:145–149.

Bezinge, A., Perreten, J., and Schafer, F. (1969). Phénomènes du lac glaciare du gorner. Technical report, Etude destinée au "Symposium on the hyrology of glaciers 1969 of the AIHS".

Blöschl, G. and Kirnbauer, R. (1992). An analysis of snow cover patterns in a small alpine catchment. *Hydrological Processes*, 6:99–109.

Blöschl, G., Kirnbauer, R., and Gutknecht, D. (1991). Distributed snowmelt simulations in an Alpine catchment, I: Model evaluation on the basis of snow cover patterns. *Water Resources Research*, 12(27):3171–3179.

Braithwaite, R. and Olesen, O. (1988). Effect of glaciers on annual runoff, johan dahl land, south greenland. *Journal of Glaciology*, 37(117):200–207.

Bruce, J. P. and Clark, R. H. (1981). *Introduction to hydrometeorology*. Pergamon Press, Oxford.

Burlando, P., Pellicciotti, F., and Strasser, U. (2002). Modelling mountainous water system between learning and speculating, looking for challenges. *Nordic Hydrology*, 33(1):47–74.

Carturan, L., Cazorzi, F., and Fontana, G. D. (2009). Enhanced estimation of glacier mass balance in unsampled areas by means of topographic data. *Annals of Glaciology*, 50(50):37–46.

Chen, J. and Ohmura, A. (1990a). Estimation of alpine glacier water resources and their change since the 1870s. In Lang, H. and Musy, A., editors, *Hydrology in Mountainous Regions*, pages 127–135. IAHS Publ. No. 193. Proceedings of two Lausanne symposia.

Chen, J. and Ohmura, A. (1990b). On the influence of alpine glaciers on runoff. In Lang, H. and Musy, A., editors, *Hydrology in Mountainous Regions*, pages 117–125. IAHS Publ. No. 193. Proceedings of two Lausanne symposia, August 1990.

Church, J., Gregory, J., Huybrechts, P., Kuhn, M., Lambeck, K., Nhuan, M., Qin, D., and Woodworth, P. (2001). Changes in sea level. In Houghton, J. and others, ., editors, *Climate Change 2001: The Scientific Basis - Contribution of Working Group I to the Third Assessment Report of the IPCC*, pages 639–694, Cambridge University Press.

Collins, I. F. (1968). On the use of the equilibrium equations and flow law in relating the surface and bed topography of glaciers and ice sheets. *Journal of Glaciology*, 7(50):199–204.

Corripio, J. G. (2004). Snow surface albedo estimation using terrestrial photography. *International Journal of Remote Sensing*, 25(24):5705–5729.

Dadic, R., Mott, R., Lehning, M., and Burlando, P. (2010). Wind influence on snow depth distribution and accumulation over glaciers. *Journal of Geophysical Research*, 115:F01012.

Davis, R. E., McKenzie, J. C., and Jordan, R. (1995). Distributed snow process modelling: an image processing approach. *Hydrological Processes*, 9:865–875.

de Saussure, H. (1880). *La question du Lac*. Genève.

Doorschot, J., Raderschall, N., and Lehning, M. (2001). Measurements and one-dimensional model calculations of snow transport over a mountain ridge. *Annals of Glaciology*, 32:153–158.

Dozier, J. and Painter, T. H. (2004). Multispectral and hyperspectral remote sensing of alpine snow properties. *Annual Review of Earth and Planetary Sciences*, 32:465–494.

Dyurgerov, M. and Meier, M. (1997). Year-to-year fluctuations of global mass balance of small glaciers and their contribution to sea-level changes. *Arctic and Alpine Research*, 29(4):392–402.

Dyurgerov, M. B. and Meier, M. F. (2002). Glacier mass balance and regime: Data of measurements and analysis. Occasional Paper 55, Institute of Arctic and Alpine Research, University of Colorado. pp. 89.

Dyurgerov, M. B. and Meier, M. F. (2005). Glaciers and the changing earth system: A 2004 snapshot. Occasional Paper 58, Institute of Arctic and Alpine Research, University of Colorado. pp. 117.

Egli, L. (2008). Spatial variability of new snow amounts derived from a dense network of alpine automatic stations. *Annals of Glaciology*, 49:51–55.

Eisen, O., Bauder, A., Lthi, M., Riesen, P., and Funk, M. (2009). Deducing the thermal structure in the tongue of Gornergletscher, Switzerland, from radar surveys and borehole measurements. *Annals of Glaciology*, 50(51):63–70(8).

Elder, K., Dozier, J., and Michaelsen, J. (1989). Spatial and temporal variation of net snow accumulation in a small alpine watershed, emerald lake basin, sierra nevada, california,u.s.a. *Annals of Glaciology*, 13:56–63.

Elder, K., Dozier, J., and Michaelsen, J. (1991). Snow accumulation and distribution in an alpine watershed. *Water Resources Research*, 27:1541–1552.

Erasov, N. V. (1968). Method to determine the volume of mountain glaciers. *Materialy Glyatsiologicheskikh Issledovanii: Khronika, Obsuzhdeniya*, 14.

Essery, R., Martin, E., Douville, H., Fernańdez, A., and Brun, E. (1999). A comparison of four snow models using observations from an alpine site. *Climate Dynamics*, 15:583–593.

Farinotti, D. (2007). Zinal-Projekt - Abschätzung der Eisvolumen, Berechnung einer Gletscherbetttopographie und Erstellung einer Abflussprognose bis ins Jahr 2100 für die Realisierung neuer Wasserfassungen der Forces Motrices de la Gougra SA. Diplomarbeit an der Versuchsanstalt für Wasserbau, Hydrologie und Glaziologie der ETH Zürich (unveröffentlicht), 198 p.

Farinotti, D., Huss, M., Bauder, A., and Funk, M. (2009a). An estimate of the glacier ice volume in the Swiss Alps. *Global and Planetary Change*, 68:225–231.

Farinotti, D., Huss, M., Bauder, A., Funk, M., and Truffer, M. (2009b). A method to estimate ice volume and ice thickness distribution of alpine glaciers. *Journal of Glaciology*, 55(191):422–430.

Farinotti, D., Magnusson, J., Huss, M., and Bauder, A. (2010). Snow accumulation distribution inferred from time-lapse photography and simple modelling. *Hydrological Processes*, 24:2087–2097.

Fliri, F. (1986). Synoptisch-klimatologische niederschlagsanalyse zwischen genfersee und hohen tauern. *Wetter Leben*, 38:140–149.

Flotron (1924-2007). Vermessung der Aaregletscher. Reports commissioned by the Kraftwerke Oberhasli (unpublished).

Forel, F. A. (1892). *Le Léman*. Slatkine Reprints, Genève. tome premier, reimprimé en 1969.

Forel, F.-A. (1895). *Les variations périodiques des glaciers.*, volume XXXIV. Archives des sciences physiques et naturelles, Genève.

Fountain, A. and Tangborn, W. (1985). The effect of glaciers on streamflow variations. *Water Resources Research*, 21(4):579–586.

Frei, C. (2007). *Climate Change and Switzerland 2050 – Impacts on Environment, Society and Economy*. Advisory Body on Climate Change (OcCC). http://www.occc.ch.

Frei, C. and Schär, C. (1998). A precipitation climatology of the Alps from high-resolution rain-gauge observations. *International Journal of Climatology*, 18(8):873–900.

Funk, M., Bösch, H., Kappenberger, G., and Müller-Lemans, H. (1997). Die Ermittlung der Eisdicke im oberen Teil des Claridenfirns (Glarner Alpen). In *Niederschlag und Wasserhaushalt im Hochgebirge der Glarner Alpen*. Schweizerische Gesellschaft für Hydrologie und Limnologie (SGHL). Beiträge zur Hydrologie der Schweiz, Nr. 36.

Funk, M., Gudmundsson, G. H., and Hermann, F. (1994). Geometry of the glacier bed of the Unteraarglacier, Bernese Alps, Switzerland. *Zeitschrift für Gletscherkunde und Glazialgeologie*, 30:187–194.

Gerecke, F. (1933). *Wellentypen, Strahlengang und Tiefenberechnung bei seismischen Eisdickenmessungen auf dem Rhonegletscher*. PhD thesis, Disseratation Gttingen. 28 p.

Glaciological Reports (1881–2008). The Swiss Glaciers, 1880–2002/03. Technical Report 1-124, Yearbooks of the Cryospheric Commission of the Swiss Academy of Sciences (SCNAT). published since 1964 by Laboratory of Hydraulics, Hydrology and Glaciology (VAW) of ETH Zürich, http://glaciology.ethz.ch/swiss-glaciers/.

Glen, J. W. (1955). The creep of polycrystalline ice. *Proceedings of the Royal Society of London, Ser. A*, 228(1175):519–538.

Good, W. and Martinec, J. (1987). Pattern recognition of air photographs for estimation of snow reserves. *Annals of Glaciology*, 9:76–80.

Gregory, J. and Oerlemans, J. (1998). Simulated future sea-level rise due to glacier melt based on regionally and seasonally resolved temperature changes. *Nature*, 391:474–476.

Gudmundsson, G. H. (1999). A three-dimensional numerical model of the confluence area of Unteraargletscher, Bernese Alps, Switzerland. *Journal of Glaciology*, 45(150):219–230.

Gudmundsson, G. H., Bauder, A., Lüthi, M., Fischer, U. H., and Funk, M. (1999). Estimating rates of basal motion and internal ice deformation from continuous tilt measurements. *Annals of Glaciology*, 28:247–252.

Gudmundsson, G. H., Thorsteinsson, T., and Raymond, C. F. (2001). Inferring bed topography and stickiness from surface data on ice streams. *Eos Trans. AGU*, 82(47). Fall Meet. Suppl., Abstract IP21A-0687.

Haeberli, W. and Hoelzle, M. (1995). Application of inventory data for estimating characteristics of and regional climate-change effects on mountain glaciers: a pilot study with the European Alps. *Annals of Glaciology*, 21:206–212.

Haeberli, W., Wächter, H., Schmid, W., and Sidler, C. (1982). Erste Erfahrungen mit dem US Geological Survey Monopuls Radioecholot im Firn, Eis und Permafrost der Schweizer Alpen. Technical report, Versuchsanstalt für Wasserbau, Hydrologie und Glaziologie der ETH Zürich. Arbeitsheft Nr. 6, 23 p.

Haefeli, R. and Kasser, P. (1948). Beobachtungen im Firn- und Ablationsgebiet des Grossen Aletschgletschers. Technical report, Versuchsanstalt für Wasserbau, Hydrologie und Glaziologie der ETH Zürich. Mitteilung Nr. 15.

Hall, D. K., Riggs, G. A., and Salomonson, V. V. (1995). Development of methods for mapping global snow cover using moderate resolution imaging spectroradiometer data. *Remote Sensing of Environment*, 54:127–140.

Helbing, J. (2005). *Glacier dynamics of Unteraargletscher: Verifying theoretical concepts through flow modeling*. Dissertation no 16303, ETH Zürich.

Hock, R. (1999). A distributed temperature-index ice- and snowmelt model including potential direct solar radiation. *Journal of Glaciology*, 45(149):101–111.

Hock, R. (2005). Glacier melt: a review of processes and their modelling. *Progress in Physical Geography*, 29(3):362–391.

Hock, R., Iken, A., and Wangler, A. (1999). Tracer experiments and borehole observations in the overdeepening of Aletschgletscher, Switzerland. *Annals of Glaciology*, 28:253–260.

Hubbard, A., Blatter, H., Nienow, P., Mair, D., and Hubbard, B. (1998). Comparison of three-dimensional model for glacier flow with field data from Haut Glacier d'Arolla, Switzerland. *Journal of Glaciology*, 44(147):368–378.

Hubbard, B., Tison, J.-L., Janssens, L., and Spiro, B. (2000). Ice-core evidence of the thickness and character of clear-facies basal ice: Glacier de Tsanfleuron, Switzerland. *Journal of Glaciology*, 46(152):140–150(11).

Huss, M. (2005). Gornergletscher - Gletscherseeausbrüche und Massenbilanzabschätzungen. Diplomarbeit an der Versuchsanstalt für Wasserbau, Hydrologie und Glaziologie der ETH Zürich (unveröffentlicht), 284 p.

Huss, M. (2009). *Past and Future Changes in Glacier Mass Balance*. PhD thesis, ETH Zürich. Dissertation No. 18230.

Huss, M. (in press). Mass balance of Pizolgletscher. *Geographica Helvetica*.

Huss, M., Bauder, A., and Funk, M. (2009a). Homogenization of long-term mass balance time series. *Annals of Glaciology*, 50(50):198–206.

Huss, M., Bauder, A., Funk, M., and Hock, R. (2008a). Determination of the seasonal mass balance of four Alpine glaciers since 1865. *Journal of Geophysical Research*, 113:F01015.

Huss, M., Farinotti, D., Bauder, A., and Funk, M. (2008b). Modelling runoff from highly glacierized alpine catchment basins in a changing climate. *Hydrological Processes*, 22(19):3888–3902.

Huss, M., Funk, M., and Ohmura, A. (2009b). Strong Alpine glacier melt in the 1940s due to enhanced solar radiation. *Geophysical Research Letters*, 36:L23501.

Huss, M., Hock, R., Bauder, A., and Funk, M. (2010a). 100-year glacier mass changes in the Swiss Alps linked to the Atlantic Multidecadal Oscillation. *Geophysical Research Letters*, 37:L10501.

Huss, M., Jouvet, G., Farinotti, D., and Bauder, A. (2010b). Future high-mountain hydrology: a new parameterization of glacier retreat. *Hydrology and Earth System Sciences Discussion*, 7:345–387.

Huss, M., Sugiyama, S., Bauder, A., and Funk, M. (2007). Retreat scenarios of unteraargletscher, switzerland, using a combined ice-flow mass-balance model. *Arctic, Antarctic and Alpine Research*, 39(3):422–431.

Huss, M., Usselmann, S., Farinotti, D., and Bauder, A. (2010c). Glacier mass balance in the south-eastern swiss alps since 1900 and perspectives for the future. *Erdkunde*, 64(2):119–140.

Hutter, K. (1983). *Theoretical glaciology; material science of ice and the mechanics of glaciers and ice sheets*. D. Reidel Publishing Company, Tokyo, Terra Scientific Publishing Company.

IPCC (1990). The IPCC scientific assessment report prepared for the Intergovernmental Panel on Climate Change, Working Group 1. Technical report, WMO/UNEP, Houghton, J.T., Jenkins, G.J. and Ephraums, J.J., eds. Cambridge University Press.

IPCC (2007). Climate change 2007: Contribution of working group i to the fourth assessment report of the ipcc. Technical report, S. Solomon, D. Qin, M. Manning, Z. Chen, M. Marquis, K.B. Averyt, M. Tignor and H.L. Miller, eds. Cambridge University Press.

Jackson, T. H. R. (1994). *A spatially distributed snowmelt-driven hydrologic model applied to Upper Sheep Creek*. Phd thesis, Utah State University, Logan, Utah.

Jansson, P., Hock, R., and Schneider, T. (2003). The concept of glacier storage - a review. *Journal of Hydrology*, 282(1–4):116–129.

Jegerlehner, J. (1902). Die Schneegrenze in den Gletschergebieten der Schweiz. In *Sonderabdruck aus Gerland's Beiträgen zur Geophysik, Band V, Heft 9*, pages 486–566.

Joerin, U., Nicolussi, K., Fischer, A., Stocker, T., and Schlüchter, C. (2008). Holocene optimum events inferred from subglacial sediments at Tschierva Glacier, Eastern Swiss Alps. *Quaternary Science Reviews*, 27:337–350.

Jóhannesson, T., Raymond, C., and Waddington, E. (1989). Time-scale for adjustment of glaciers to changes in mass balance. *Journal of Glaciology*, 35(121):355–369.

Jost, W. (1936). Die seismischen Eisdickenmessungen am Rhonegletscher 1931. Technical report, Denkschriften der Schweizerischen Naturforschenden Gesellschaft. Band 71, Abh. 2, 42 p. mit Kartenbeilage.

Jouvet, G., Huss, M., Picasso, M., Rappaz, J., and Blatter, H. (2009). Numerical simulation of rhone's glacier from 1874 to 2100. *Journal of Computational Physics*, 228(17):6426–6439.

Jouvet, G., Picasso, M., Rappaz, J., and Blatter, H. (2008). A new algorithm to simulate the dynamics of a glacier: theory and applications. *Journal of Glaciology*, 54(188):801–811.

Kamb, B. and Echelmeyer, K. A. (1986). Stress-gradient coupling in glacier flow: I. longitudinal averaging of the influence of ice thickness and surface slope. *Journal of Glaciology*, 32(111):267–284.

Kasser, P. (1973). Influence of changes in the glacierized area on summer run-off in the Porte du Scex drainage basin of the rhone. In *Symposium on the Hydrology of Glaciers*, pages 221–225, Cambridge, 7-13 September 1969. IAHS Publ. No. 95.

Kasser, P. and Aellen, M. (1974). Die Gletscher der Schweizer Alpen 1970/71. Technical report, Gletscherkommission der Schweizerischen Naturforschenden Gesellschaft. 92. Bericht.

Kasser, P. and Aellen, M. (1976). Die Gletscher der Schweizer Alpen 1971/72 und 1972/73. Technical report, Gletscherkommission der Schweizerischen Naturforschenden Gesellschaft. 93. und 94. Bericht.

Kasser, P., Aellen, M., and Siegenthaler, H. (1982). Die Gletscher der Schweizer Alpen 1973/74 und 1974/75. Technical report, Gletscherkommission der Schweizerischen Naturforschenden Gesellschaft. 95. und 96. Bericht.

Kasser, P., Aellen, M., and Siegenthaler, H. (1983). Die Gletscher der Schweizer Alpen 1975/76 und 1976/77. Technical report, Gletscherkommission der Schweizerischen Naturforschenden Gesellschaft. 97. und 98. Bericht.

Kind, R. J. (1981). Snow drifting. In Gray, D. M. and Male, D. H., editors, *Handbook of snow: principles, processes, management and use*, pages 338–359. Elsevier, New York.

König, M., Winther, J.-G., and Isaksson, E. (2001). Measuring snow and glacier ice properties from satellite. *Reviews of Geophysics*, 39(1):1–27.

Kreis, A. (1944). Seismische Sondierungen auf dem Morteratschgletscher. Technical report, Verhandlungen der Schweizerischen Naturforschenden Gesellschaft. p. 95.

Kreis, A., Süsstrunk, A., and Florin, F. (1947). Bericht ber die Seismische Sondierungen auf dem Glacier de la Plaine Morte. (unveröffentlicht), 7 p. mit Kartenbeilage.

Lang, H. (1987). Forecasting meltwater runoff from snow-covered areas and from glacier basins. In Kraijenhoff, D. and Moll, J., editors, *River Flow Modelling and Forecasting*, pages 99–127, Dordrecht. Reidel.

Leavesley, G. H. and Stannard, L. G. (1990). Application of remotely sensed data in distributed-parameter watershed model. In Kite, G. W. and Wankiewicz, A., editors, *Proceedings of the workshop on applications of remote sensing in hydrology*, pages 47–64, Saskatoon.

Lehning, M., Loewe, H., Ryser, M., and Raderschall, N. (2008). Inhomogeneous precipitation distribution and snow transport in steep terrain. *Water Resources Research*, 44(7):W07404.

Lehning, M., Völksch, I., Gustafsson, D., Nguyen, T., Staehli, M., and Zappa, M. (2006). Alpine3d: a detailed model of mountain surface processes and its application to snow hydrology. *Hydrological Processes*, 20:2111–2128.

Liston, G. E. and Sturm, M. (1998). A snow-transport model for complex terrain. *Journal of Glaciology*, 44(148):498–516.

Lliboutry, L. A. (1979). Local frictions laws for glaciers: a critical review and new openings. *Journal of Glaciology*, 23(89):67–95.

Luce, C. H., Tarboton, D. G., and Cooley, K. R. (1998). The influence of the spatial distribution of snow on basin-averaged snowmelt. *Hydrological Processes*, 12:1671–1683.

Luckman, B. H. (1977). The geomorphic activity of snow avalanches. *Geografiska Annaler*, 59(1/2):31–48.

Macheret, Y. Y., Cherkasov, P. A., and Bobrova, L. I. (1988). Tolschina i ob'em lednikov djungarskogo alatau po danniy aeroradiozondirovaniya. *Materialy Glyatsiologicheskikh Issledovanii: Khronika, Obsuzhdeniya*, 62:59–71.

Macheret, Y. Y. and Zhuravlev, A. B. (1982). Radio-echo-sounding of svalbard glaciers. *Journal of Glaciology*, 28(99):295–314.

Machguth, H., Eisen, O., Paul, F., and Hoelzle, M. (2006). Strong spatial variability of snow accumulation observed with helicopter-borne gpr on two adjacent alpine glaciers. *Geophysical Research Letters*, 33:L13503.

Maisch, M., Wipf, A., Denneler, B., Battaglia, J., and Benz, C. (2000). Die Gletscher der Schweizer Alpen. vdf Hochschulverlag AG, ETH Zürich. Schlussbericht NFP31.

Martinec, J. and Rango, A. (1981). Areal distribution of snow water equivalent evaluated by snow cover monitoring. *Water Resources Research*, 17(5):1480–1488.

Meier, M. (1984). Contribution of small glaciers to global sea level. *Science*, 226:1418–1421.

Meier, M., Dyurgerov, M., Rick, U., O'Neel, S., Pfeffer, W., Anderson, R., Anderson, S., and Glazovsky, A. (2007). Glaciers dominate eustatic sea-level rise in the 21st century. *Science*, 317:1064–1067.

Mercanton, P. (1936). Les sondages seismométriques de la commission des glaciers á l'Unteraar. Technical report, Actes de la Société Helvetique des Sciences Naturelles. p. 271-273.

Mercanton, P. L. (1958). Aires englacées et cotes frontales des glaciers suisses - Leurs changements de 1876 á 1934 d'aprés l'Atlas Siegfried et la Carte Nationale et quelques indications sur les variations de 1934 á 1957. *Couru d'eau et énergie*, 12.

Mesinger, F. and Pierrehumbert, R. T. (1986). Alpine lee cyclogenesis: Numerical simulation and theory. In *Scientific Results of the Alpine Experiment (ALPEX), Volume I*, pages 141–163. GARP Publication Series. No. 27.

Mittaz, C., Imhof, M., Hoelzle, M., and Haeberli, W. (2002). Snowmelt evolution mapping using an energy balance approach over an alpine terrain. *Arctic, Antarctic and Alpine Research*, 34(3):274–281.

Mothes, H. (1927). Seismische Dickenmessungen von Gletschereis. *Zeitschrift für Geophysik*, 3:121–134.

Mothes, H. (1929). Neue ergebnisse der eiseismik. *Journal of Geophysics*, pages 120–144.

Müller, B. (2004). Veränderung am Triftgletscher seit 1861: Untersuchung der Stabilitätsveränderung der Steilstufe. Diplomarbeit an der VAW/ETH-Zürich, (unveröffentlicht).

Müller, F., Caflisch, T., and Müller, G. (1976). Firn und Eis der Schweizer Alpen: Gletscherinventar. Publ. Nr. 57, Geographisches Institut der ETH Zürich, Zürich.

Müller, F., Caflisch, T., and Müller, G. (1977). Instructions for compilation and assemblage of data for a world glacier inventory. Technical report, International Commission on Snow and Ice - Temporary Technical Secretariat for World Glacier Inventory, Departement of Geography, ETH Zurich.

Nemoto, M. and Nishimura, K. (2004). Numerical simulation of snow saltation and suspension in a turbulent boundary layer. *Journal of Geophysical Research*, 109:D18206.

Nye, J. F. (1965). The flow of a glacier in a channel of rectangular, elliptic or parabolic cross-section. *Journal of Glaciology*, 5(41):661–690.

Oberholzer, P. and Salamí, F. (1998). Geophysikalische Untersuchungen der Eisverhältnisse am Rossbodengletscher (Simplongebiet, Wallis). Diplomarbeit an der VAW-ETH Zürich, (unveröffentlicht).

Oeschger, H., Schotterer, U., Stauffer, B., Haeberli, W., and Röthlisberger, H. (1977). First results from alpine core drilling projects. *Zeitschrift für Gletscherkunde und Glazialgeologie*, 13:193–208. Heft 1/2.

Østrem, G. (1973). Runoff forecasts for highly glacierized basins: The role of snow and ice in hydrology. In *Proceedings of the Banff Symposium*, pages 1111–1129. IAHS Publ. No. 107.

Paterson, W. S. B. (1970). The application of ice physics to glacier studies. In Demers, J., editor, *Glaciers. Proceedings of the Workshop Seminar, September 24 and 25, 1970*, pages 43–46. Ottawa, Secretariat, Canadian National Committee for the International Hydrological Decade.

Paterson, W. S. B. (1994). *The Physics of Glaciers*. Pergamon, New York, third edition.

Paul, F. (2004). *The new Swiss glacier inventory 2000 - Application of remote sensing and GIS*. PhD thesis, Department of Geography, University of Zurich. 198 pp.

Peck, E. L. and Brown, M. J. (1962). An approach to the development of isohyetal maps for mountainous areas. *Journal of Geophysical Research*, 67:681–694.

Radić, V. and Hock, R. (2010). Regional and global volumes of glaciers derived from statistical upscaling of glacier inventory data. *Journal of Geophysical Research*, 115:F01010.

Radić, V., Hock, R., and Oerlemans, J. (2007). Volume-area scaling vs flowline modelling in glacier volume projections. *Annals of Glaciology*, 46:234–240.

Raper, S. C. B. and Braithwaite, R. J. (2005). The potential for sea level rise: New estimates for glacier and ice cap area and volume distributions. *Geophysical Research Letters*, 32:L05502.

Raper, S. C. B. and Braithwaite, R. J. (2006). Low sea level rise projections from mountain glaciers and icecaps under global warming. *Nature*, 439:311–313.

Raymond, M. (2007). *Estimating basal properties of glaciers and ice streams from surface measurements*. Dissertation no 17362, ETH Zürich.

Renaud, A. and Mercanton, P. (1948). Les sondages sismiques de la Commission Helvetique des glaciers. Technical report, Bureau Central Seismologique International. Série A, Travaux Scientifiques, fasc. 17, p. 66-78.

Riesen, P. (2006). Auswertung der Radarmessungen auf dem Gornergletscher vom April 2005: Vergleich zweier Methoden. Semesterarbait an der Versuchsanstalt für Wasserbau, Hydrologie und Glaziologie der ETH Zürich (unveröffentlicht), 70 p.

Rohrer, M. (1989). Determination of the transition air temperature from snow to rain and intensity of precipitation. In Sveruk, B., editor, *Precipitation measurement*, pages 475–482. ETH Zurich, Switzerland.

Röthlisberger, H. (1979). Glaziologische Arbeiten im Zusammenhang mit den Seeausbrüchen am Grubengletscher, Gemeinde Saas Balen, Wallis. Technical report, Versuchsanstalt für Wasserbau, Hydrologie und Glaziologie der ETH Zürich. Mitteilung Nr. 41, p. 233-256.

Röthlisberger, H. and Lang, H. (1987). Glacial hydrology. In Gurnell, A. and Clark, M., editors, *Glacio-fluvial Sediment Transfer*, pages 207–284, New York. Wiley.

Röthlisberger, H. and Vögtli, K. (1976). Recent d.c. resistivity soundings on Swiss glaciers. *Journal of Glaciology*, 6(47):607–621.

Schäfli, B., Hingray, B., Niggli, M., and Musy, A. (2005). A conceptual glacio-hydrological model for high mountainous catchments. *Hydrology and Earth System Sciences*, 9:95–109.

Schmidt, R. A. (1986). Transport rate of drifting snow and the mean wind speed profile. *Boundary Layer Meteorology*, 34:213–241.

Schneeberger, C., Blatter, H., Abe-Ouchi, A., and Wild, M. (2003). Modelling changes in the mass balance of glaciers of the northern hemisphere for a transient $2 \times CO_2$ scenario. *Journal of Hydrology*, 282(1-4):145–163.

Schotterer, U., Haeberli, W., Good, W., Oeschger, H., and Röthlisberger, H. (1981). Datierung von kaltem Firn und Eis in einem Bohrloch vom Colle Gnifetti, Monte Rosa. In *Gletscher und Klima*, pages 170–212. Jahrbuch der Schweizerischen Naturforschenden Gesellschaft. wissenschaftlicher Teil, 1978.

Schuler, T., Fischer, U. H., Sterr, R., Hock, R., and Gudmundsson, G. H. (2002). Comparison of modeled water input and measured discharge prior to a release event: Unteraargletscher, Bernese Alps, Switzerland. *Nordic Hydrology*, 33(1):27–46.

Seyfried, M. S. and Wilcox, B. P. (1995). Scale and the nature of spatial variability: Field examples having implications for hydrologic modeling. *Water Resources Research*, 31:173–184.

Sharp, M. J., Richards, K., Willis, I., Arnold, N., Nienow, P., Larson, W., and Tison, J.-L. (1993). Geometry, bed topography and drainage system structure of the Haut Glacier d'Arolla, Switzerland. *Earth Surface Processes and Landforms*, 18:557–572.

Shi, J. and Dozier, J. (2000a). Estimation of snow water equivalence using sir-c/x-sar, part i: Inferring snow density and subsurface properties. *IEEE Transactions on geoscience and remote sensing*, 38(6):2465–2474.

Shi, J. and Dozier, J. (2000b). Estimation of snow water equivalence using sir-c/x-sar, part ii: Inferring snow depth and particle size. *IEEE Transactions on geoscience and remote sensing*, 38(6):2475–2488.

Solomon, S. and others, . (2007). Climate change 2007: the physical science basis. Technical report, Contribution of Working Group I to the Fourth Assessment Report of the Intergovernmental Panel on Climate Change. Cambridge University Press.

Spreen, W. C. (1947). A determination of the effect of topography upon precipitation. *Transactions of the American Geophysical Union*, 2:285–290.

Sugiyama, S., Bauder, A., Funk, M., and Zahno, C. (2007). Evolution of Rhonegletscher, Switzerland, over the past 125 years and in the future: application of an improved flowline model. *Annals of Glaciology*, 46:268–274.

Sugiyama, S., Bauder, A., Huss, M., Riesen, P., and Funk, M. (2008). Triggering and drainage mechanisms of the 2004 glacier-dammed lake outburst in Gornergletscher, Switzerland. *Journal of Geophysical Research*, 113:F04019.

Süsstrunk, A. (1950). Seismische Messungen auf Gletschern (1948-50). Technical report, Verhandlungen der Schweizerischen Naturforschenden Gesellschaft.

Süsstrunk, A. (1951). Sondages du glacier par la méthode sismique. *La Houille Blanche*, pages 309–319. Numéro spécial A/1951.

Süsstrunk, A. (1959). Seismische Untersuchungen auf dem Findelengletscher, Bericht an die Grande Dixance SA, Sion. (unveröffentlicht), 7 p. mit Kartenbeilage.

Swissborung (1958-1959). Berichte ber die thermischen und mechanischen Sondierungen auf dem Findelengletscher an die Grande Dixance SA, Sion.

Tarboton, D. G., Chowdhury, T. G., and Jackson, T. H. (1995). A spatially distributed energy balance snowmelt model. In Tonnessen, K. A., Williams, M. W., and Tranter, M., editors, *Biogeochemistry of seasonally snow-covered catchments, Proceedings of a Boulder Symposium*, pages 141–155. IAHS Publ. No. 228.

Thorsteinsson, T., Raymond, C. F., Gudmundsson, G. H., Bindschadler, R. A., Vornberger, P., and Joughin, I. (2003). Bed topography and lubrication inferred from surface measurements on fast-flowing ice streams. *Journal of Glaciology*, 49(167).

Thyssen, F. (1979). Bericht über die Eisdickenmessungen auf dem Colle Gnifetti, Monte Rosa. (unveröffentlicht), 12 p.

Thyssen, F. and Ahmad, M. (1969). Ergebnisse seismischer Messungen auf dem Aletschgletscher. *Polarforschung*, 39(1):283–293.

Turpin, O., Ferguson, R., and Johansson, B. (1999). Use of remote sensing to test and update simulated snow cover in hydrological models. *Hydrological Processes*, 13:2067–2077.

UNESCO/IAHS (1970). Perennial ice and snow masses. *Technical Papers in Hydrology*, 1.

VAW (1983a). Bericht No. 52.22 (Allalingletscher) an die Kraftwerke Mattmark SA, Saas Grund (VS). Technical report, Versuchsanstalt für Wasserbau, Hydrologie und Glaziologie der ETH Zürich. 28 p.

VAW (1983b). Bericht No. 59.23 (Griesgletscher) an die Kraftwerke Aegina SA, Ulrichen (VS). Technical report, Versuchsanstalt für Wasserbau, Hydrologie und Glaziologie der ETH Zürich.

VAW (1998). Mauvoisin – Giétrogletscher – Corbassièregletscher. Glaziologische Studien im Zusammenhang mit den Stauanlagen Mauvoisin. Im Auftrag der Elektrizitätsgesellschaft Lauffenburg AG.

VAW (2002). Radarmessungen Oberer Theodulgletscher. Technical report, Versuchsanstalt für Wasserbau, Hydrologie und Glaziologie der ETH Zürich. Bericht Nr. 7943.52.01 im Auftrag von B. Schnyder, Büro für angewandte Glaziologie, Saas Fee.

VAW (2004). Glaziologische Untersuchungen am Triftgletscher im Zusammenhang mit dem Rückzug der Gletscherzunge, der Bildung eines proglazialen Sees und der Gefahr von gefährlichen Eisabbrüchen in den See. Bericht Nr. 7943b.20.01 im Auftrag vom Oberingenieurkreis 1 des Kantons Bern.

VAW (2006). Glacier de Zinal - Rapport sur la prise d'eau prévue sous le Glacier de Zinal. Technical report, Versuchsanstalt für Wasserbau, Hydrologie und Glaziologie der ETH Zürich. Expertise mandatée par les Forces Motrices de la Gougra SA.

VAW (2007a). Glacier de Zinal - Rapport sur les prises d'eau prévues dans le bassin versant de Zinal - Résultats de l'étude glaciologique et hydrologique. Technical report, Versuchsanstalt für Wasserbau, Hydrologie und Glaziologie der ETH Zürich. Expertise mandatée par les Forces Motrices de la Gougra SA.

VAW (2007b). Unterer Grindelwaldgletscher - Glaziologische Abklrungen im Zusammenhang mit der Seebildung. Technical report, Versuchsanstalt für Wasserbau, Hydrologie und Glaziologie der ETH Zürich. Bericht Nr. 7945.21.2 im Auftrag des Oberingenieurkreis I des Kantons Bern.

VAW (2009). Eisvolumen der Gletscher im im Mattmarkgebiet. Technical report, Versuchsanstalt für Wasserbau, Hydrologie und Glaziologie der ETH Zürich. Bericht Nr. 7902.52.55 im Auftrag der Kraftwerke Mattmark AG.

VAWE (1961). Bericht No. 59.5 an die Motor Columbus AG, Baden. Technical report, Versuchsanstalt für Wasserbau und Erdbau der ETH Zürich. 6 p. mit Kartenbeilage.

Vieli, A., Funk, M., and Blatter, H. (1997). Griesgletscher: Berechnungen des Gletscherfliessens und Perspektiven für die Zukunft. *Wasser, Energie, Luft*, 89(5/6):107–114.

von Deschwanden, A. (2008). Eisvolumenbestimmung des Griesgletschers und die Beziehung zwischen Eisvolumen und Gletscherflächet - auswertung von radarmessungen 1999. BsC-thesis at VAW - ETH Zurich, unpublished, 46 p.

Wächter, H. (1979). Eisdickenmessungen auf dem Rhonegletscher - Ein Versuch mit Radio-Echo Sounding. Diplomarbait am Geographischen Institut der ETH Zürich (unveröffentlicht), 67 p.

Wallén, C. C., editor (1970). *Climates of Northern and Western Europe*. World Survey of Climatology, Vol. 5 and 6. Elsevier Scientific Publishing Company, Amsterdam.

Weertman, J. (1964). The theory of glacier sliding. *Journal of Glaciology*, 5(39):287–303.

Witmer, U. (1984). Eine Methode zur flächendeckenden Kartierung von Schneehöhen unter Berücksichtigung von relievbedingten Einflüssen. *Geographica Bernensia*, 21. Geographisches Institut der Universität Bern.

Zahno, C. (2004). Der Rhonegletscher in Raum und Zeit: Neue geometrische und klimatische Einsichten. Diplomarbeit an der VAV - ETH Zürich (unveröffentlicht), 132 p.

Zemp, M., Haeberli, W., Hoelzle, M., and Paul, F. (2006). Alpine glaciers to disappear within decades? *Geophysical Research Letters*, 33(13):L13504.

Zemp, M., Hoelzle, M., and Haeberli, W. (2009). Six decades of glacier mass-balance observations: a review of the worldwide monitoring network. *Annals of Glaciology*, 50(50):101–111.

Zhurovlyev, A. B. (1985). Korrelyatsionniy metod ozenky zapasov l'da v lednikakh. *Materialy Glyatsiologicheskikh Issledovanii: Khronika, Obsuzhdeniya*, 52:241–249.

Zweifel, B. (2000). Kernbohrungen in kalten gletschern. MS-thesis at ETH Zurich, unpublished.

Die VDM Verlagsservicegesellschaft sucht für wissenschaftliche Verlage abgeschlossene und herausragende

Dissertationen, Habilitationen, Diplomarbeiten, Master Theses, Magisterarbeiten usw.

für die kostenlose Publikation als Fachbuch.

Sie verfügen über eine Arbeit, die hohen inhaltlichen und formalen Ansprüchen genügt, und haben Interesse an einer honorarvergüteten Publikation?

Dann senden Sie bitte erste Informationen über sich und Ihre Arbeit per Email an *info@vdm-vsg.de*.

Sie erhalten kurzfristig unser Feedback!

VDM Verlagsservicegesellschaft mbH
Dudweiler Landstr. 99
D - 66123 Saarbrücken

Telefon +49 681 3720 174
Fax   +49 681 3720 1749

www.vdm-vsg.de

Die VDM Verlagsservicegesellschaft mbH vertritt

Printed by Books on Demand GmbH, Norderstedt / Germany